PROCEEDINGS OF THE DUBNA CONFERENCE ON THE MÖSSBAUER EFFECT

1963

Authorized translation from the Russian
by
Springer Science+Business Media, LLC

ISBN 978-1-4899-4850-2 ISBN 978-1-4899-4848-9 (eBook)
DOI 10.1007/978-1-4899-4848-9

Softcover reprint of the hardcover 1st edition 1995

FOREWORD

The Working Conference on the Mössbauer Effect was held in Dubna (USSR), July 3-6, 1962. About 60 physicists from the regional divisions of the Joint Institute of Nuclear Studies participated. The conference was organized by the Laboratory of Neutron Physics of the Joint Institute, which, in addition to neutron investigations, also carries out research in other fields of nuclear physics that deal with the Mössbauer Effect.

This collection contains the papers presented to the Organizing Committee of this conference.

Two review articles by Yu. M. Kagan on the theory of the Mössbauer Effect and its application to solid state physics and local fields in nuclei have been omitted from this collection. The content of these articles was almost identical to that of the introductory article in the collection "The Mössbauer Effect," published in 1962 by the Foreign Literature Press.

V. P. Alfimenkov and A. V. Strelkov were responsible for the organization and publication of these papers.

CONTENTS

APOLOGIA

So that the following 24 papers might be made available in English translation to participants in the Third International Conference on the Mössbauer Effect, their translation has been expedited, and they appear here exactly as provided by the translators, without benefit of the usual editorial refinements.

We regret the inconsistencies of terminology inevitable under such conditions, and regret also the occasional rendition of the Russian prefix сверх- in its more common meaning of "super-," rather than as "hyper-," as is correct in the word "hyperfine."

We take this opportunity to express our great appreciation to the contributing translators, whose unstinted cooperation made it possible for us to provide this English version to the Third International Conference in a period measured in weeks. We extend the very best wishes of the Consultants Bureau staff for a highly successful meeting.

Frances Coleman
Editor
Consultants Bureau
Enterprises, Inc.

PROBLEMS OF METHODOLOGY-
NEW MÖSSBAUER EMITTERS

METHODS AND COMPILATION OF DATA ON MOSSBAUER γ-TRANSITIONS

V. P. Alfimenkov, Yu. M. Ostanevich

Up to the present time, about 150 original papers have appeared, devoted to the experimental study of recoilless emission and resonance absorption of γ-quanta by nuclei [Mössbauer effect /1/] and its various applications. This rich material makes it possible to single out the most important methods used, which will be discussed in this paper.

I. SOURCES, FILTERS, AND DETECTORS

The first question that experimenters are faced with is making up suitable sources for solving their problems. Depending on the actual problem, the source strengths used vary between very wide limits from 50 μCu /2/ to 0.4 Cu /3/. In observing the Mössbauer effect, the source must necessarily be a solid. In getting the radioactive nuclei into the material to be used, wide use is made of thermal diffusion, carrier precipitation from solution, and co-electrolysis. Often the compound in which the source is subsequently to be used is irradiated directly. High temperature annealing is used to remove possible crystal lattice defects. There is a wide variety of compounds into which the radioactive nuclei can be put. Which compounds are best must be found empirically in each particular case. However we can give some general rules here.

Sources with the smallest unsplit emission line width are found by using paramagnetic metals or alloys with a cubic lattice. This is due to the small electron spin relaxation times in paramagnetic metals ($<10^{-11}$ sec) /4,5/.

Another rule deals with the probability P of recoilless γ emission. This probability is usually appreciably higher in compounds like oxides, where the strong bond between the emitting nucleus and the light atoms gives a large amount of importance to the optical branches in the frequency spectrum of the lattice /6/. However, using oxides as sources is not always the best thing to do. Often the crystal lattice of the oxide does not have high enough symmetry, and the emission line is broadened or split by the intracrystalline field /7--10/. Oxides are not recommended in cases where the γ-transition under study is preceded by K-capture or a strongly converged γ- transition.

It has been shown in /8/, and confirmed experimentally in /11/ that in these cases the Auger effect can cause the radiating nucleus to have a considerable probability of being in a distorted electron environment. As a result, the emission line may be shifted or broadened, which complicates the spectrum in a way that is difficult to interpret. It should however be noted that this phenomenon has not been observed in the work with $Sn^{119}O_2$ /12/.

The spectra of the compounds used as filters are considerably richer, although the above rules remain in force. A very promising step has been taken recently, namely, that of using polymers /13/, and organic compounds /14--16/ as filters.

As far as method is concerned, all the work that has been done may be divided into two groups (Fig. 1). The experiments in the first group were made in transmission geometry. In the second group of experiments, what is detected is secondary radiation, such as scattered γ-quanta, conversion electrons /17,18/, or the X-rays emitted after conversion /19,20/. The second group of experiments includes detecting the γ-activity induced in the filter by resonance absorption of γ- quanta. This method, proposed in /21/ is only convenient for very narrow lines, and has not yet been used. The methods of the second type are generally marked by a gain in magnitude of effect observed, with a considerable loss in light intensity.

The usual soft γ spectroscopy detectors are, as a rule, used in work on the Mössbauer effect, including luminescent screens containing thin NaI (Tl) crystals, and proportional counters with various filling materials. The magnitude of the effect observed depends to a large extent on the type and quality of the detector used, since the γ- spectrum under study almost always has γ- transitions or X-rays in it as a background.

A communication /22/ has been presented at the conference on a Geiger counter with a high selective sensitivity to the γ- quanta emitted without recoil by Sn^{119} nuclei. This detector greatly facilitates work on Sn^{119}, but it is evidently impossible to set up this type of detector for the majority of the Mössbauer γ- transitions being studied.

II. METHODS OF VARYING THE γ ENERGY, AND RECORDING SYSTEMS.

In work on the Mossbauer effect, it is necessary to investigate the absorption of γ- radiation from the source by the filter as a function of the relative energy shift between the emission and absorption spectra.

The longitudinal Doppler effect is used as the basic method of producing a controllable shift between the emission and absorption spectra /23/. Moving the source with respect to the filter (or vice versa) produces an energy shift $\Delta E = E_0\ (v/c)$ where E_0 is the γ-transition energy, and v is the rate of motion.

In practice, the most diverse systems are used to produce the relative motion, and they may be classified as follows:

1) Mechanical devices such as wheels, dollys, cams, levers, and other mechanisms.

2) Electromagnetic vibrators of various types. Dynamic loud speakers are very widely used because of their simplicity.

3) Hydraulic devices and piezoelectric crystals are used at small and very small velocities (10^{-1}--10^{-5} cm/sec).

In order to get the energy picture of resonance absorption (spectrum), the rate of the relative motion, which fixes the energy shift, must be variable. The following basic methods of changing the velocity with time may be cited (Fig. 2):

1) Motion at a constant velocity, produced by wheels /23--25/. The velocity spectrum is obtained from successive changes in the effect at different rates of motion. This method is the most promising one for getting very high velocities, but it gives poor velocity resolution (5--10%).

2) Back and forward motion, with the velocity constant for a certain length of time. Produced by various types of vibrators: mechanical /7,9,16,26--34/, electromagnetic /48--51/, and piezoelectric /8/. In this method, either one point is measured, or two points on the spectrum symmetric about zero velocity. The whole spectrum is obtained by successive changes in velocity. This

method is widely used because the recording systems are simple, and the velocities are accurately known.

3) Periodic motion at a continuously changing velocity. Produced by cams and electromagnetic vibrators. Two velocity change laws are used (Fig. 6): linear and harmonic. The first case gives the results of the measurements in more convenient form than the second. The linear law is produced by either mechanical /26/, or electromagnetic /4,52/ vibrators. The harmonic law is produced by electromagnetic vibrators /3,4,9,18,20, 40--47/, which is particularly simple.

The counters are constructed to suit the law of motion. One or two counting channels are usually used when making measurements at a constant velocity. All the points in the spectrum can be got through fast enough so that changes in counting rate from drift in the apparatus have no effect on the velocity spectrum. Occasionally, part of the time is spent on a regular counting rate measurement with an obviously resolved resonance /8/.

Two different recording methods are possible when working with a continuously changing velocity. The first of these is no different from the one just discussed, but instead of a constant velocity, the mean over a period is used. This method /40,41,55/ permits getting information with the simplest possible equipment on the magnitude of the effect and the upper limit of line width. However, it has not been widely used, because of the very poor velocity resolution. Another, basic method consists of using multichannel amplitude analyzers in conjunction with modulating devices.

When the pulse comes from the detector, the modulator forms a pulse with an amplitude proportional to the modulating voltage, which is recorded by the multichannel amplitude analyzer. The modulating voltage of special form is phased in a definite way with the periodic relative motion, so that each instantaneous value of the velocity corresponds with a definite analyzer channel. A sawtooth or triangular /26,52/ modulating voltage is used with a linear velocity change law. If the velocity change is harmonic, a sinusoidal modulating voltage /42--44/ is also used in addition to a linear modulating voltage /9,20,39,45,46/. Quite often, the modulating voltage is obtained from a pickup directly connected to the moving element of the vibrator.

There is also interest in systems where the recording is done directly in the memory circuit of the analyzer, without the need for a double time--amplitude--time transformation /18,47,57/.

The principal value of multichannel systems (in addition to convenience in measurement) is that the whole spectrum is taken at one time, which reduces the effect of counting rate drift in the equipment to a minimum. Nevertheless "single channel" systems have one serious advantage: if necessary, they permit making a very accurate study of some particular portion of the spectrum without losing time going over the whole spectrum.

Usually, no difficulties arise in measuring the velocities when the velocity of the motion is constant. It is more difficult to measure the velocities when they are changing continuously or in jumps, i.e., when there is some kind of oscillatory motion. Calculated velocities are often used, obtained from the amplitude of the oscillation and the law of motion /18,26,33,37/. However, this is only correct if the law of motion is quite accurately known. A stroboscope in conjunction with a microscope /57/ has been used to find the law of motion of a vibrator under operating conditions. In a number of cases, a split Mössbauer spectrum where the distance between the components is known has been used to calibrate the velocity scale. The nuclear Zeeman effects /4/ and frequency modulation /8/ were used to get the spectrum. This last method is of wider applicability, but has so far not been extensively used because of technical difficulties.

III. TREATMENT OF EXPERIMENTAL RESULTS.

In working up the results of the measurements, account usually has to be taken of the background due to quanta of other energies or to X-rays. The magnitude of the background may be found by the usual spectroscopic methods, since the Mössbauer effect presents nothing specific in this respect. It should only be noted that if $\sigma_{res} < \sigma_{at}$, and the γ- spectrum is complicated, finding the probabilities (P, P') of recoilless γ emission and absorption may require measuring the background to a quite high degree of accuracy. The magnitude of the resonance absorption is found from the relative difference in counting rates when resonance exists and when it does not. Measuring these two quantities does not present any difficulties.

The form of the line, and the way it depends on the thickness of the source and the filter have been investigated in /58--60/. In the last of these papers, a study was also made of the form of a spectrum consisting of two lines. In short, the results of these studies are as follows. It is primarily the broadening of the lines observed that depends on the thickness of the source and the filter. The form begins to undergo appreciable changes only at very large thicknesses. In a poorly resolved spectrum, the true distances between the components may be appreciably different from those observed. The experimental line shape may be written either as an integral, or as a series of Bessel functions of imaginary argument, which permits complicated spectra to be analyzed on a machine /61/.

A possible source of broadening of the observed lines produced by the equipment is the finite velocity resolution coming either from uncontrollable deviations from the desired law of motion, or from the finite angular divergence of the beam. These questions have not yet been investigated from a quantitative angle. The recoilless γ emission and absorption probabilities P and P' are found from measurements made with filters of various thicknesses /58,60, 62/. Two methods are possible. If the velocity spectrum contains only one line, P and P' may be found from the way in which the maximum effect depends on the filter thickness. These measurements are particularly simple for undisplaced spectra. Another way is to find how the area under the absorption curve varies with filter thickness. This method is more laborious, but it is also more widely applicable. It can be used for broadened and split lines.

No particular experimental difficulties are presented by measuring the shifts of various sorts after a technique has been developed for finding the velocity spectra. Especial attention must be given to getting a reliable calibration of the velocity scale and fixing the energy shift zero. This point may always be found experimentally by measuring the spectrum when the source and filter are identical. As we have said before, bad errors are possible when using amplitude modulation as a result of inaccurate phasing. This danger may be avoided by using scanning over the whole period of the motion. By this method, the displacement between the source and filter lines may be found independently of the position of the apparatus zero /57/.

Measuring the splitting for well split components does not present any particular difficulties either, with the exception of the case where the distance between the components is much greater than the line width. Incomplete splitting requires a careful approach, and is discussed in /60/.

IV. COMPILATION OF RESULTS AND DISCUSSION OF OUTLOOK.

Up to the present time, the Mössbauer effect has been investigated with more or less success on 22 γ- transitions (Table 1). Of these, four: Fe^{57}, Sn^{119}, Au^{197}, Du^{161}, are used for various "applied" studies. For the rest of the transitions, studies have been made to some degree or other of the properties of the excited level, and of those compounds in which they were studied.

In addition to those given in Table 1, about 60 γ-transitions are known which on first glance appear to be suitable for investigating the Mössbauer effect. In order to evaluate how real the possibilities are of investigating these transitions, we have introduced the parameter $P^2(0)\ \sigma_{res}/\sigma_{at}$, where P(0) is the probability of recoilless γ emission at 0^oK. In calculating P(0), the Debye approximation was used for a one-atom cubic lattice, together with the Debye temperature for the pure elements. That the proposed parameter gives the best measure of the possibilities of making an experiment may be debatable, but, in any case, this quantity is proportional to the ease with which a given transition may be investigated.

It may be seen from Table 1 that, with rare exceptions, γ-transitions with the parameter less than 0.1 have not been investigated. There are only three exceptions:

1) Zn^{67}. Studies entail considerable difficulties.

2) Pr^{141}. No effect observed with an accuracy of $\pm 0.2\%$.

3) Ta^{181}. No data, simply the information that an effect has been observed /60/.

Table 2 gives the γ-transitions of the nuclei for which our parameter is greater than 0.1, for which there are suitable mother nuclei, and for which a satisfactory γ- spectrum is to be expected. The nucleus Ge^{73} is included in the table for other considerations. Here apparently the investigations run into considerable experimental difficulties. Nevertheless, the narrow line, and the possibilities of application in semiconductor physics may serve to justify further investigation. We feel that the conclusion may be drawn from a comparison of the two tables that the possibilities of investigating new Mossbauer γ radiators are far from exhausted at the present time. The investigation of new Mössbauer radiators is of great interest both in nuclear physics and from an applied standpoint.

We do not feel that the table given exhausts all possibilities. As further improvements are made in technique, the experimental possibilities will unquestionably be extended.

United Institute of Nuclear Studies

REFERENCES

/1/ R. L. Mössbauer, UFN 72, No. 4, 658 (1960).

/2/ R. V. Pound, G. A. Rebka, Jr., Phys. Rev. Lett. 3, 554 (1959)

/3/ R. V. Pound, G. A. Rebka, Jr., Phys. Rev. Lett. 4, 337 (1960)

/4/ G. K. Wertheim, Phys. Rev. 121, 63 (1961)

/5/ W. Low, Proc. VII-th Confer. on Low Temper. Phys. (1960)
/6/ Yu. Kagan, ZhÉTF, 41 659 and 1296 (1961)
/7/ V. V. Sklvarevskii et.al., ZhÉTF, 40, 1874 (1961)
/8/ V. P. Alfimenkov et.al., ZhÉTF, 42, 1029 (1962)
/9/ M. Kalvins, Zs. Naturforsch. 17a, 248 (1962)
/10/ O. C. Kistner, A. W. Sunyar. Phys. Rev. Lett. 4, 412 (1960)
/11/ G. K. Wertheim, Phys. Rev. 124, 764 (1961)
/12/ N. N. Delyagin et.al. ZhÉTF 41, 1347 (1961)
/13/ V. A. Bryukhanov et.al. ZhÉTF 42, 637 (1962)
/14/ A. Yu. Aleksandrov et.al., Report to the Conference.
/15/ V. I. Gol'danskii et.al., Report to the Conference.
/16/ Zahn, U. a.o. Zs. Phys. 166, 220 (1961)
/17/ K. P. Mitrofanov, V. S. Shpinel', ZhÉTF 40, 983 (1961)
/18/ E. Kankelkeit, Zs. Phys. 164, 442 (1961)
/19/ Pham Zuy Xien et.al. ZhÉTF 42, 703 (1962)
/20/ H. Frauenfelder a.o. N. Cim. 19, 183 (1961)
/21/ S. I. Aksenov et.al. ZhÉTF 40, 88 (1961)
/22/ K. P. Mitrofanov et.al., Report at this conference
/23/ R. L. Mossbauer, Zs. Naturforsch, 14a, 211 (1959)
/24/ G. De Pasquali a.o. Phys. Rev. Lett. 4, 71 (1960)
/25/ D. A. Shirley a.o. Phys. Rev. 123, 816 (1961)
/26/ V. A. Bryukhanov et.al. PTÉ, 1962, No. 1, 23
/27/ L. Grodzins, F. Genovese, Phys. Rev. 121, 228 (1961)
/28/ F. E. Obenshaim, H. Wegener, Phys. Rev. 121, 1344 (1961)
/29/ S. Ofer a.o. Phys. Rev. 120, 406 (1960)
/30/ S. Hanna a.o. Phys. Rev. Lett. 4, 177 (1960)
/31/ Yu. Bara et.al. Nukleonika 7, 135 (1962)
/32/ O. I. Sumbaev et.al. ZhÉTF 42, 115 (1962)

/33/ R. L. Mössbauer, W. H. Widemann, Zs. Phys. 159, 33 (1960)

/34/ J. G. Dash a.o. Phys. Rev. 122, 1116 (1961)

/35/ R. Bauminger, S. G. Cohen, a.o. Phys. Rev. 122, 743 (1961)

/36/ A.T.F. Boyle a.o. Proc. Phys. Soc. 79, 416 (1962)

/37/ A.T.F. Boyle a.o. Proc. Phys. Soc. 77, 1062 (1961)

/38/ A.T.F. Boyle a.o. Proc. Phys. Soc. 77, 129 (1961)

/39/ P. I. Black, P. B. Moon, Nature. 188, 482 (1960)

/40/ S. Iha a.o. Nuovo Cim. 19, 682 (1961)

/41/ M. Cordey-Hayes a.o. Proc. Phys. Soc. 75, 810 (1960)

/42/ S. L. Ruby a.o. Zs. Phys. 31, 580 (1960)

/43/ W. Kerler a.o. Zs. Phys. 167, 176 (1962)

/44/ O. C. Kistner a.o. Phys. Rev. 123, 179 (1961)

/46/ M. Kalvins a.o. Zs. Phys. 163, 87 (1961)

/47/ M. Kalvins a.o. Zs. f. Naturforsch. 17a, 494 (1962)

/48/ C. Alf, G. K. Wertheim, Phys. Rev. 122, 1415 (1961)

/49/ C. E. Johnson, Phys. Rev. Lett. 6, 450 (1961)

/50/ F. T. Lynch a.o. Phys. Rev. 120, 513 (1960)

/51/ G. K. Wertheim, Phys. Rev. Lett. 4, 403 (1960)

/52/ G. K. Wertheim a.o. Phys. Rev. 125

/53/ O. C. Kistner, A. W. Sunyar, Phys. Rev. Lett. 4, 412 (1960)

/54/ R. V. Pound, G. A. Rebka, Jr., Phys. Rev. Lett. 4, 275 (1960)

/55/ R. V. Pound, G. A. Rebka, Jr., Phys. Rev. Lett. 4, 397 (1960)

/56/ V. P. Alfimenkov et.al. ZhÉTF 42, 1036 (1962)

/57/ V. P. Alfimenkov et.al., Report at this conference

/58/ S. Margulies, J. R. Ehrman, Nucl. Instr. 12, 131 (1961)

/59/ S. L. Ruby, J. M. Hicks, Rev. Sci. Instr. 33, No. 1, 27,(1961)

/60/ G. A. Bykov and Pham Zuy Hien, Report at this conference

/61/ R. Bauminger a.o. Phys. Rev. Lett. 6, 467 (1961)

/62/ P. P. Craig a.o. Phy. Rev. Lett. 3, 221 (1959)

/63/ H. Frauenfelder, The Mössbauer Effect. New York, 1962

Review from the series "Frontiers in Physics".

Translated by Chas. V. Larrick

Nucleus	% in natural mixture	E_γ, keV	$T_{1/2}$, sec	α, conversion coeff.	P^2 σ_0/σ_{atom}
Fe^{57}	2,17	14,4	$1{,}0\cdot 10^{-7}$	15	200
Ni^{61}	1,25	75	$5{,}2\cdot 10^{-9}$	K 0,11	55
Zn^{67}	4,11	93	$9{,}4\cdot 10^{-6}$	K 0,63	$2{,}7\cdot 10^{-2}$
Sn^{119}	8,58	24	$1{,}9\cdot 10^{-8}$	7,3	180
Te^{125}	6,99	35	$1{,}6\cdot 10^{-9}$	K 12	54
Pr^{141}	100	145	$2\cdot 10^{-9}$	0,39	$1{,}3\cdot 10^{-6}$
Sm^{149}	13,8	22,5	$>4\cdot 10^{-9}$	30	5,8
Eu^{153}	52,23	103	$3{,}4\cdot 10^{-9}$	4,3	0,14
Gd^{155}	14,7	87		0,4	0,25
Dy^{161}	11,8	25,7	$2{,}8\cdot 10^{-8}$	0,22	240
Er^{166}	33,4	80	$1{,}8\cdot 10^{-9}$	1,7	2.8
Tm^{169}	100	8,4	$4{,}0\cdot 10^{-9}$		
Yb^{170}	3,03	84,2	$1{,}57\cdot 10^{-9}$	1,6	1,85
Hf^{177}	18,5	113	$4{,}2\cdot 10^{-10}$	0,75	0,12
Ta^{181}	100	136,1	$5{,}7\cdot 10^{-11}$	1,5	$2{,}4\cdot 10^{-2}$
W^{182}	26,4	100	$1{,}3\cdot 10^{-9}$	4,5	6
W^{183}	14.4	46,5	$5{,}7\cdot 10^{-10}$	9	10,6
W^{186}	14,4	99,1	$1{,}5\cdot 10^{-10}$	3,5	3,5
R^{187}	62,93	134	$2\cdot 10^{-9}$	2,1	0,18
Ir^{191}	38,5	129	$1{,}4\cdot 10^{-10}$	2,9	0,20
Ir^{193}	61,5	73	$5{,}7\cdot 10^{-9}$	1,5	24
Au^{197}	100	77	$1{,}9\cdot 10^{-9}$	2,5	2,4

Nucleus	% in natural mixture	E_γ, keV	$T_{1/2}$, sec	α, conversion coeff.	$P^2(0)$ σ_0/σ_{atom}
Ge^{73}	7,67	13,5	$3{,}1\cdot 10^{-6}$	3600	$7{,}3\cdot 10^{-2}$
I^{127}	100	59		1.9	1,5
Xe^{129}	26,4	40	$7\cdot 10^{-10}$	22*	8
Eu^{151}	47,8	22		12	0,5
Gd^{155}	14,73	60		4*	0,6
Gd^{156}	20,47	89	$2\cdot 10^{-9}$	1	1,4
Tb^{159}	100	58	$3{,}5\cdot 10^{-11}$	6	3,2
Dy^{160}	2,29	87	$1{,}8\cdot 10^{-9}$	1,5	2,5
Er^{168}	27,07	79,8	1,8·10[illegible]	2,1	2,0
Yb^{171}	14,31	66,7	$<5\cdot 10^{-7}$	6	1,5
Yb^{173}	16,13	78,7		6	0,3
Hf^{180}	35,22	93	$1{,}4\cdot 10^{-9}$	4	1,6
W^{180}	0,135	102		5	3,9
W^{184}	30,6	111	$1{,}3\cdot 10^{-9}$	0,8*	7
Re^{185}	37,07	125		2,4	0,3
Os^{186}	1,59	137	$5{,}1\cdot 10^{-10}$	0,45	0,8
Pt^{195}	33,8	99		9	0,8
Pt^{195}	33,8	129		1,2*	0,18
Hg^{201}	13,22	32,1		70*	1,5

* Calculated Values

Fig. 1.

I

II

S — Source
K — Collimator
P — Absorber
D — Detector

Shield Interrupter Shield

Fig. 2.

$V(t)$	Recording	Modulating wave form	Spectrum observed	Ways of doing
	Multi-channel ampl. analyzer		N~ No. of pulses n — channel no.	A. B. Parabola
$V_0 = V_0 \cos \omega t$	Multi-channel ampl. analyzer	u Without modulation	$\Delta E = 0$	Or at resonance

Fig. 3.

$V(t)$	Recording	Ways of doing
$V_0 = const$	One counting channel	$V_0 = V \cos \alpha$ S — Source P — Absorber M — Motor D — Detector
$V = +V_0 \quad 0 < t < \frac{T}{2}$ $V = -V_0 \quad \frac{T}{2} < t < T$	One or two counting channels	A. B. C. D.

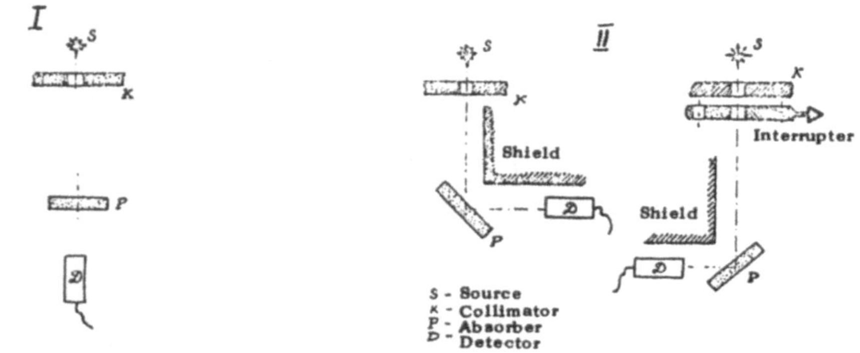

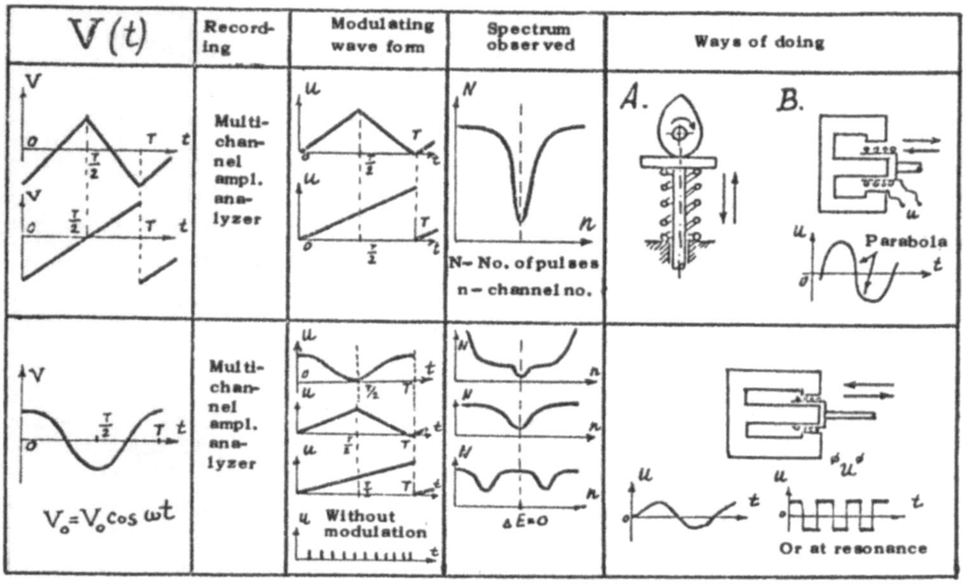

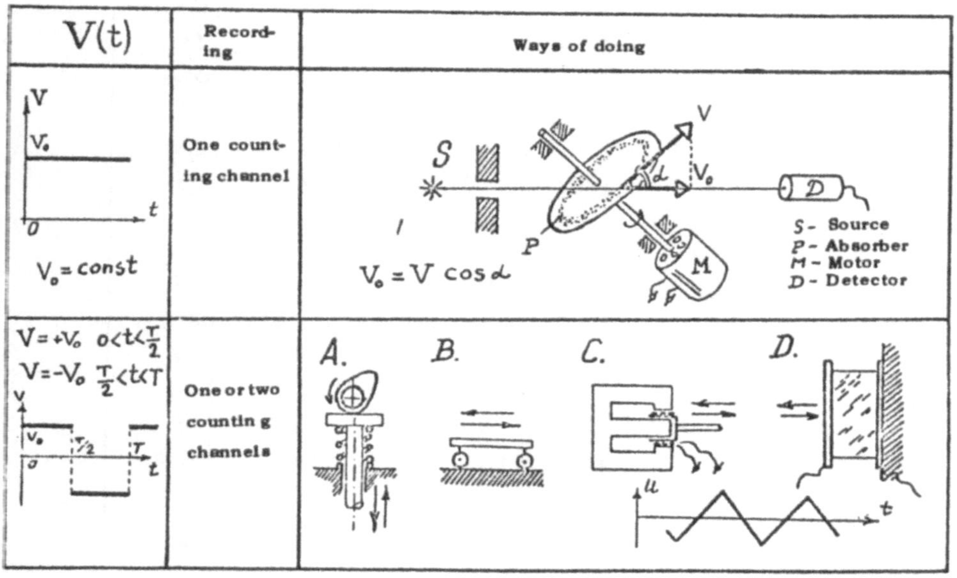

CALCULATION OF THE PARAMETERS OF AN EXPERIMENTAL SPECTRUM OF THE RESONANT ABSORPTION OF γ-QUANTA IN CRYSTALS

G. A. Bykov and Pham Zuy Hien

ANNOTATION

The parameters of an observed spectrum of the resonant absorption of γ-quanta in crystals are calculated for single and split lines, with allowance for self-absorption in the source.

1. INTRODUCTION

In experiments on the resonant absorption of γ-quanta in crystals one investigates the dependence of the magnitude of the absorption on the relative velocity of the source and absorber. We have to deal, to a greater or lesser extent, with the problem of allowing for some associated factors (self-absorption in the source, absorber thickness, etc.) on the measured quantities. Usually in these investigations the necessary corrections are calculated numerically on the basis of certain assumptions. However, the limits of applicability of the assumptions made are often ill-defined.

Shirley, Kaplan and Axel /2/ gave a series of empirical formulas for the parameters of the observed spectrum, based on the results of a numerical calculation by Margulies and Ehrman /1/. Some general formulas for the calculation of the necessary experimental parameters are derived below.

2. STATEMENT OF THE PROBLEM

The shape of the spectrum in experiments on the resonant absorption of γ-quanta from a moving source is given by the quantity

$$\mathcal{E}(v) = \frac{N(\infty) - N(v)}{N(\infty)}, \tag{1}$$

where v is the relative velocity of the source and absorber, N(v) is the rate of counting of γ-quanta which have passed through the resonant absorber moving at the velocity v and $N(\infty)$ is the corresponding rate for the absorber moving at a sufficiently high velocity so that there is no resonant absorption.

Let E be the energy of a γ-quantum; $\sigma(E)$, the effective cross section for the resonant absorption; $W_e(E)$, the energy distribution function for the emission spectrum, normalized to unit area; n_A, the number of atoms of the investigated nuclide per 1 cm^2 of the absorber; E_0, the energy of the resonance level; and Γ, the total width of the level; then, using the nondimensional variables $x = (E - E_0)/(\Gamma/2)$ and $y = (E_0 v/c)/(\Gamma/2)$, we can write for $\varepsilon(y)$ (cf. /3/):

$$\mathcal{E}(y) = \alpha f \int_{-\infty}^{+\infty} \left[1 - e^{\sigma(x) \cdot n_A}\right] W_e(x+y)\, dx \tag{2}$$

Here f is the probability of the recoil-free emission of a γ-quantum, and α is the relative proportion of resonant quanta in the emission spectrum.

It follows from Eq.(2) that the area under the experimental absorption spectrum $S_{exp} = \int_{-\infty}^{\infty} \varepsilon(y)dy$ is independent of the emission line shape and is governed only by the absorption spectrum. In particular S_{exp} is independent of self-absorption in the source and of stray vibrations of the apparatus. For this reason S_{exp} is a convenient experimental parameter.

For a single absorption line $\sigma(x)$ is given by the formula:

$$n_A \sigma(x) = \frac{\sigma_0 f' n_A}{1+x^2}, \tag{3}$$

where f' is the probability of the recoil-free absorption of a γ-quantum, and

$$\sigma_0 = \frac{2I_b+1}{2I_a+1} \frac{\lambda^2}{2\pi} \frac{\Gamma_\gamma}{\Gamma}.$$

Here I_a, I_b are the spins of the ground and excited states of the nucleus, λ is the wavelength of the γ-quanta, and Γ_γ is the radiation width of the level. We shall denote the nondimensional quantity $\sigma_0 f' n_A$ by C_A.

If the absorption line is split into p components, then $\sigma(x)$ is written in the form

$$n_A \sigma(x) = \sum_{i=1}^{p} \frac{C_{iA}}{1+(x+\Delta_i)^2} \tag{4}$$

where Δ_i is the shift of the i-th component relative to some fixed energy level, for example the energy of the unsplit line, in units of $\Gamma/2$; $C_{iA} = \beta_i C_A$, where β_i is the relative intensity of the i-th component.

The calculation given below is based on the relationships (2), (3), (4).

3. SINGLE EMISSION AND ABSORPTION LINES

a) Source with no self-absorption.

For such a source $W_e(x)$ has the form

$$W_e(x) = \frac{1}{\pi} \frac{1}{1+x^2} \tag{5}$$

Combining Eqs. (2), (3) and (5) we arrive at the following expression for $\varepsilon(y)$

$$\mathcal{E}(y) = \alpha f \left\{ \frac{1}{\pi} \int_{-\infty}^{+\infty} \left[1 - \exp\left(- \frac{C_A}{1+x^2} \right) \right] \frac{1}{1+(x+y)^2} dx \right\} . \tag{6}$$

The value of the maximum absorption $\varepsilon(0)$, calculated in /4/, is

$$\mathcal{E}(0) = \alpha f \left[1 - e^{-C_A/2} I_0(C_A/2) \right] , \tag{7}$$

where $I_0(A)$ is a Bessel function of zero order with an imaginary argument. The other parameters of the distribution $\varepsilon(y)$ are calculated as follows. For S_{exp} it follows from Eq. (6):

$$S_{exp} = \int_{-\infty}^{+\infty} \mathcal{E}(y)\, dy = \alpha f \int_{-\infty}^{+\infty} \left(1 - e^{-\frac{C_A}{1+x^2}} \right) dx . \tag{8}$$

The integral in Eq. (8) is calculated exactly (cf. /5/)

$$\int_{-\infty}^{+\infty} \left(1 - e^{-\frac{C_A}{1+x^2}} \right) dx = \pi K(C_A) , \tag{9}$$

where the function $K(A)$ is defined as follows:

$$K(\lambda) = \lambda e^{-\lambda/2} \left[I_0\left(\frac{\lambda}{2}\right) + I_1\left(\frac{\lambda}{2}\right) \right] , \tag{10}$$

and $I_1(\lambda)$ is a first-order Bessel function with an imaginary argument. Finally we have

$$S_{\text{эксп}} = \alpha f \pi C_A e^{-\frac{C_A}{2}} \left[I_0\left(\frac{C_A}{2}\right) + I_1\left(\frac{C_A}{2}\right) \right] . \tag{11}$$

Since $K(\lambda)$ represents the processes considered here, its behavior in the limiting cases should be noted:

$$\frac{K(\lambda)}{\lambda} \to 1 \text{ as } \lambda \to 0 , \qquad \frac{K(\lambda)}{\sqrt{\lambda}} \to \frac{2}{\sqrt{\pi}} \text{ as } \lambda \to \infty . \tag{12}$$

The quantities $\varepsilon(0)$ and S_{exp} allow us to represent the shape of the absorption spectrum. For this purpose we shall consider the integrated width of the distribution $\varepsilon(y)$:

$$\varkappa_{\text{integ}}(C_A) = \frac{1}{\mathcal{E}(0)} \int_{-\infty}^{+\infty} \mathcal{E}(y)\, dy . \tag{13}$$

We note that $\varkappa_{integ}(C_A)$ represents the distribution to the same extent as the more frequently used distribution width at half-height, $2\varkappa(C_A)$, where $\varkappa(C_A)$ is defined as

$$\mathcal{E}[\varkappa(C_A)] = \frac{1}{2}\mathcal{E}(0). \tag{14}$$

Eqs. (7) and (11) lead to the relationship

$$\varkappa_{integ}(C_A) = \pi \frac{\varkappa(C_A)}{1 - e^{-C_A/2} I_0(C_A/2)}. \tag{15}$$

It should be noted that at low values of C_A it follows from Eq. (6) that

$$\mathcal{E}(y) = \frac{\alpha f}{\pi} C_A \frac{2}{2^2 + y^2} \quad \text{for} \quad C_A \to 0. \tag{16}$$

Eq. (16) suggests that in a known range of variation of C_A the quantity $\varepsilon(y)$ can be approximated by a dispersion distribution which, taking into account Eq. (11), is written as

$$\mathcal{E}(y) = \alpha f \frac{\varkappa(C_A)\,\varkappa(C_A)}{[\varkappa(C_A)]^2 + y^2}. \tag{17}$$

In this case $\varkappa(C_A)$ is related in a simple way to $\varkappa_{integ}(C_A)$:

$$\varkappa_{integ}(C_A) = \pi \kappa(C_A). \tag{18}$$

Hence using Eq. (14) we obtain

$$\varkappa(C_A) = \kappa(C_A) \frac{1}{1 - e^{-C_A/2} I_0(C_A/2)}. \tag{19}$$

In the limiting cases we have

$$\varkappa(C_A) \to 2 \text{ as } C_A \to 0; \quad \varkappa(C_A) \to 2\sqrt{\frac{C_A}{\pi}} \text{ as } C_A \to \infty. \tag{20}$$

To find the region of applicability of Eq. (19) we shall compare $\varkappa(C_A)$ with the numerically calculated data given graphically in /1/. When $0 \leqslant C_A \leqslant 6$ the agreement is exact; when $6 \leqslant C_A \leqslant 10$ deviations are found reaching 3% when $C_A = 10$. Hence we see that in the range of variation of C_A which is important in practice, the observed distribution has the dispersion form of Eq. (17). Appendix I gives a formal proof of this statement.

b) <u>Source with self-absorption.</u>

If the radiating nuclei are uniformly distributed throughout the source thickness, then the emission spectrum is given by

$$W_e(x) = \frac{1}{N(C_s)}\left[1 - \exp\left(-\frac{C_s}{1+x^2}\right)\right]. \tag{21}$$

Here $C_s = \sigma_0 f n_s$, n_s is the number of nuclei of the resonantly absorbing nuclide per 1 cm^2 of the source, and the normalization constant is $N(C_s) = \int_{-\infty}^{\infty}\left(1 - e^{-\frac{C_s}{1+x^2}}\right)dx$. From Eq. (9) we have

$$N(C_s) = \pi K(C_s).$$

The experimental spectrum is then

$$\mathcal{E}(y) = \frac{\alpha f}{\pi K(C_s)}\int_{-\infty}^{+\infty}\left[1 - \exp\left(-\frac{C_A}{1+x^2}\right)\right]\left[1 - \exp\left(-\frac{C_s}{1+(x+y)^2}\right)\right]dx \tag{22}$$

S_{exp} is, as before, given by Eq. (11). The maximum absorption is given by

$$\mathcal{E}(0) = \frac{\alpha f}{\pi K(C_s)}\int_{-\infty}^{+\infty}\left[1 - \exp\left(-\frac{C_A}{1+x^2}\right)\right]\left[1 - \exp\left(1 - \frac{C_s}{1+x^2}\right)\right]dx. \tag{23}$$

The integral in Eq. (22) is calculated exactly and we find that

$$\mathcal{E}(0) = \frac{\alpha f}{K(C_A)}\left[K(C_A) + K(C_s) - K(C_A + C_s)\right]. \tag{24}$$

Using the formula $\lambda \frac{d}{d\lambda} I_p(\lambda) + p I_p(\lambda) = \lambda I_{p-1}(\lambda)$ [1], it is easy to obtain the relationship $\mathcal{E}(0) = \alpha f\left[1 - e^{-C_A/2} I_0\left(\frac{C_A}{2}\right)\right]$ as $C_s \to 0$. In work on the resonant absorption of γ-quanta in crystals one determines the important physical parameter f' from the experimental

---

[1] For the properties of Bessel functions used here and later see /6/.

dependence of $\varepsilon(0)$ on C_A. Figure 1 shows that it is essential to allow for self-absorption in the source. The quantity $\varepsilon(0)$ is found to be the most sensitive parameter of the observed distribution to variation in the shape of the emission spectrum. Therefore to determine f' it is usually preferable to employ the dependence of the integrated absorption on C_A, especially as from the experimental point of view the S_{exp} curve is more convenient (cf. Fig. 2 for $\Delta = 0$).

We shall assume that in a certain known range of variation of the quantities C_A and C_s the observed spectrum has the form

$$\mathcal{E}(y) = \alpha f \frac{\kappa(C_A)\varkappa(C_A, C_s)}{[\varkappa(C_A, C_s)]^2 + y^2} . \tag{24}$$

For the half-width we obtain the formula

$$\varkappa(C_A, C_s) = \frac{\kappa(C_A)\kappa(C_s)}{\kappa(C_A) + \kappa(C_s) - \kappa(C_A + C_s)} , \tag{25}$$

from which we obtain the relationship (19) as $C_s \to 0$. It is useful to bear in mind that $\varkappa(C_A, C_s) = \varkappa(C_s, C_A)$. Comparison of Eq. (25) with the numerical calculations in /1/ shows discrepancies not greater than 3%. Hence we can draw the following conclusions:

1. For a wide range of variation of the parameters C_A, C_s (at least for $C_A, C_s \leqslant 10$) the experimental absorption spectrum has the dispersion form of Eq. (24) with a half-width given by Eq. (25). Thus we see that on going over from $C_A, C_s \to 0$ to finite values of C_A, C_s only the width of the observed spectrum changes substantially, while the shape of the distribution $\varepsilon(y)$ undergoes only a slight change.

2. With an accuracy sufficient for later calculations we obtain the equality

$$\int_{-\infty}^{+\infty}\left(1-e^{-\frac{C_A}{1+x^2}}\right)\left(1-e^{-\frac{C_S}{1+(x+y)^2}}\right)dx = \pi\frac{K(C_S)K(C_A)\varkappa(C_A,C_S)}{[\varkappa(C_A,C_S)]^2+y^2} \qquad (26)$$

4. SPLITTING OF EMISSION AND ABSORPTION LINES

As a result of the interaction of the electric and magnetic moments of the nuclei with external or internal (crystal) fields, the emission and absorption lines are often split into separate components. The investigation of split lines is of great interest, since they may provide information on the magnetic moment and quadrupole electric moment of the nuclei in the excited state as well as information on the internal crystal fields.

If the observed spectrum is in the form of separate completely split components (the difference of the energies of the nearest components being $\gg \Gamma$), then the results of the preceding section can be applied to each component of the true absorption or emission spectrum. The case when there is "partial" splitting requires special consideration. Below we outline the general procedure for calculating the required parameters using as an example the doublet splitting of an absorption line.

Let us assume that

$$n_A\sigma(x) = \frac{C_{1A}}{1+x^2} + \frac{C_{2A}}{1+(x+\Delta)^2}, \qquad (27)$$

where Δ is the energy difference between the components, in units of $\Gamma/2$. For the emission spectrum we employ the distribution of Eq. (21), which allows for self-absorption in the source, and therefore for the observed splitting we have (the subscript "2" denotes splitting)

$$\varepsilon_2(y) = \frac{\alpha f}{\pi K(C_s)} \int_{-\infty}^{+\infty} \left[1 - \exp\left(-\frac{C_{1A}}{1+x^2} - \frac{C_{2A}}{1+(x+\Delta)^2}\right)\right]\left[1 - \exp\left(-\frac{C_s}{1+(x+y)^2}\right)\right] dx \,. \tag{28}$$

$S_{2.\exp}$ is given by the expression

$$S_{2\exp} = \alpha f \int_{-\infty}^{+\infty} \left[1 - \exp\left(-\frac{C_{1A}}{1+x^2} - \frac{C_{2A}}{1+(x+\Delta)^2}\right)\right] dx \,, \tag{29}$$

which is easily calculated from Eqs. (9) and (26). Finally we obtain the formula

$$S_{2\exp} = \pi \alpha f \left\{ K(C_{1A}) + K(C_{2A}) - \frac{\varkappa(C_{1A})\varkappa(C_{2A})\varkappa(C_{1A}, C_{2A})}{[\varkappa(C_{1A}, C_{2A})]^2 + \Delta^2} \right\} . \tag{30}$$

Hence in the limiting cases of "small" and "large" splitting we obtain the results:

1. $S_{2\exp} = \pi \alpha f K(C_{1A} + C_{2A}) = S_{\text{эксп}}(C_{1A} + C_{2A}) \,, \quad \Delta \to 0 \,;$
2. $S_{2\exp} = \pi \alpha f [K(C_{1A}) + K(C_{2A})] = S_{\text{эксп}}(C_{1A}) + S_{\text{эксп}}(C_{2A}) \,, \quad \Delta \to \infty \,.$

In the case of identical components $(C_{1A} = C_{2A})$ the dependence of $S_{2.\exp}$ on C_A is as given in Fig. 2.

To calculate the other parameters we have to find the function $\varepsilon_2(y)$ in an explicit form. We shall transform Eq. (28) in the following way

$$\varepsilon_2(y) = \frac{\alpha f}{\pi K(C_s)} [J_1 + J_2 - J_3] \,, \tag{28^{I}}$$

where J_1, J_2, J_3 are the following integrals

$$J_1 = \int_{-\infty}^{+\infty} \left[1 - \exp\left(-\frac{C_{1A}}{1+x^2}\right)\right]\left[1 - \exp\left(-\frac{C_s}{1+(x+y)^2}\right)\right] dx \,,$$

$$J_2 = \int_{-\infty}^{+\infty} \left[1 - \exp\left(-\frac{C_{2A}}{1+x^2}\right)\right]\left[1 - \exp\left(-\frac{C_s}{1+(x+y)^2}\right)\right] dx \,, \tag{28^{II}}$$

$$J_3 = \int_{-\infty}^{+\infty} \left[1 - \exp\left(-\frac{C_{1A}}{1+x^2}\right)\right]\left[1 - \exp\left(-\frac{C_{2A}}{1+(x+\Delta)^2}\right)\right]\left[1 - \exp\left(-\frac{C_s}{1+(x+y)^2}\right)\right] dx \,.$$

J_1, J_2 represent the parts of the absorption due to the "independent" action of the components, and J_3 is the part of the absorption due to the mutual overlap of the components. From Eq. (26) it follows that

$$\mathcal{J}_1 = \pi \frac{\varkappa(C_{1A})\, \varkappa(C_{1A}, C_s) . \varkappa(C_s)}{[\varkappa(C_{1A}, C_s)]^2 + y^2} \; ; \; \mathcal{J}_2 = \pi \frac{\varkappa(C_{2A})\, \varkappa(C_s)\, \varkappa(C_{2A} . C_s)}{[\varkappa(C_{2A}, C_s)]^2 + (y-\Delta)^2} \; . \qquad (28^{III})$$

To calculate the integral $\mathcal{J}_3$ we shall make use of symmetry properties:

$$\begin{aligned}
&1.\; \mathcal{J}_3(C_{1A}, C_{2A}, C_s; \Delta, y) = \mathcal{J}_3(C_{1A}, C_{2A}, C_s; -\Delta, -y)\\
&2.\; \mathcal{J}_3(C_{1A}, C_{2A}, C_s; \Delta, y) = \mathcal{J}_3(C_{1A}, C_s, C_{2A}; y, \Delta)\\
&3.\; \mathcal{J}_3(C_{1A}, C_{2A}, C_s; \Delta, y) = \mathcal{J}_3(C_s, C_{2A}, C_{1A}; y-\Delta, y) \qquad (31)\\
&4.\; \mathcal{J}_3(C_{1A}, C_{2A}, C_s; \Delta, y) = \mathcal{J}_3(C_{2A}, C_{1A}, C_s; \Delta, \Delta-y) .
\end{aligned}$$

We note also the following relationships:

$$\begin{aligned}
&1.\; \int_{+\infty}^{+\infty} \mathcal{J}_3(C_{1A}, C_{2A}, C_s; \Delta, y)\, dy = \pi^2 \varkappa(C_s) \frac{\varkappa(C_{1A})\varkappa(C_{2A})\varkappa(C_{1A}, C_{2A})}{[\varkappa(C_{1A}, C_{2A})]^2 + \Delta^2}\\
&2.\; \mathcal{J}_3(C_{1A}, C_{2A}, C_s; \Delta, y) = \pi \Big\{ \frac{\varkappa(C_{1A})\varkappa(C_s)\varkappa(C_{1A}, C_s)}{[\varkappa(C_{1A}, C_s)]^2 + y^2} + \qquad (32)\\
&\quad + \frac{\varkappa(C_{2A})\varkappa(C_s)\varkappa(C_{2A}, C_s)}{[\varkappa(C_{2A}, C_s)]^2 + y^2} - \frac{\varkappa(C_{1A}+C_{2A})\varkappa(C_s)\varkappa(C_{1A}+C_{2A}, C_s)}{[\varkappa(C_{1A}+C_{2A}, C_s)]^2 + y^2} \Big\} .
\end{aligned}$$

The relationships (31), (32) are useful for checking the correctness of the approximate formulas for $\mathcal{J}_3$ and also for determining the parameters in these formulas.

The method given below for the calculation of $\mathcal{J}_3$ is based on the integral Fourier transformation /7/. We shall denote the function $1 - \exp[-\lambda/(1 + x^2)]$ by $\mathcal{J}_o(\lambda, x)$ and its Fourier transform by $\mathcal{J}_o^*(\lambda, w)$. $\mathcal{J}_o(\lambda, x)$ and $\mathcal{J}_o^*(\lambda, w)$ are related by

$$\begin{aligned}
\mathcal{J}_o(\lambda, x) &= \frac{1}{2\pi} \int_{-\infty}^{+\infty} \mathcal{J}_o^*(\lambda, w)\, e^{iwx}\, dw ,\\
\mathcal{J}_o(\lambda, w) &= \int_{-\infty}^{+\infty} \mathcal{J}_o(\lambda, x)\, e^{-iwx}\, dx . \qquad (33)
\end{aligned}$$

Then Eq. (28^{II}) can be rewritten in the form

$$\begin{aligned}
\mathcal{J}_3(C_{1A}, C_{2A}, C_s; \Delta, y) &= \int_{-\infty}^{+\infty} \mathcal{J}_o(C_{1A}, x)\mathcal{J}_o(C_{2A}, x+\Delta)\mathcal{J}_o(C_s, x+y)\, dx = \qquad (34)\\
&= \int_{-\infty}^{+\infty} \mathcal{J}_o(C_{1A}, x)\mathcal{J}_o(C_{2A}, \Delta-x)\mathcal{J}_o(C_s, x-y)\, dx .
\end{aligned}$$

From Eq. (33) we have

$$\mathcal{J}_o(C_s, x-y) = \frac{1}{2\pi}\int_{-\infty}^{+\infty}\mathcal{J}_o^*(C_s,\omega)\,e^{i(x-y)\omega}\,d\omega;$$
$$\mathcal{J}_o(C_{2A}, \Delta-x) = \frac{1}{2\pi}\int_{-\infty}^{+\infty}\mathcal{J}_o^*(C_{2A},\omega')\,e^{i(\Delta-x)\omega'}\,d\omega'. \qquad (33^{I})$$

Substitution of Eq. (33^{I}) into (34) gives

$$\mathcal{J}_3(C_{1A},C_{2A},C_s,\Delta,y) = \left(\frac{1}{2\pi}\right)^2\int_{-\infty}^{+\infty}\mathcal{J}_o^*(C_s,\omega)\,\mathcal{J}_o^*(C_{2A},\omega')\,e^{i(\Delta\omega'-y\omega)}\,d\omega\,d\omega'\int_{-\infty}^{+\infty}\mathcal{J}_o(C_{1A},x)\,e^{ix(\omega-\omega')}\,dx. \qquad (34^{I})$$

Since from Eq. (33) $\int_{-\infty}^{+\infty}\mathcal{J}_o(C_{1A},x)e^{-ix(\omega'-\omega)}dx = \mathcal{J}_o^*(C_{1A},\omega'-\omega)$, we obtain the result

$$\mathcal{J}_3(C_{1A},C_{2A},C_s,\Delta,y) = \left(\frac{1}{2\pi}\right)^2\int_{-\infty}^{+\infty}\mathcal{J}_o^*(C_s,\omega)\,e^{-iy\omega}\int_{-\infty}^{+\infty}e^{i\Delta\omega'}\mathcal{J}_o^*(C_{2A},\omega')\,\mathcal{J}_o^*(C_{1A},\omega'-\omega)\,d\omega\,d\omega'. \qquad (34^{II})$$

Eq. (34) reduces the problem of finding an approximate expression for $\mathcal{J}_3$ to the selection of a reliable approximation for the Fourier transformation of $\mathcal{J}_o(\lambda, x)$. For this purpose we shall return to Eq. (26). Applying the Fourier transformation to $\mathcal{J}_o(C_A, x)$, $\mathcal{J}_o(C_s, y-x)$ and the dispersion part of the curve (26) we obtain the equality

$$\mathcal{J}_o^*(C_A,\omega).\mathcal{J}_o(C_s,\omega) = \pi^2 K(C_A).K(C_s)\,e^{-\varkappa(C_A,C_s)|\omega|}. \qquad (35)$$ [2)]

Since Eq. (35) is valid over a wide range of variation of the parameters, we obtain

---

2) Eqs. (35) and (34) are a special case of a general theory of convolution-type functions and their Fourier transformations. It is known /6/ that the Fourier transformation of a convolution of two functions is equal to the product of their Fourier transformations, and a Fourier transformation of a product of two functions is a convolution of the Fourier transformations of these functions. An example of Fourier transformations used in physical applications and leading to convolution-type relations arises in the problem of the distorting effect of spectroscopic instruments /8/.

$$\mathcal{J}_0^*(C_A, \omega) = \pi K(C_A) e^{-\varkappa(C_A)|\omega|} \; ; \quad \mathcal{J}_0^*(C_s, \omega) = \pi K(C_s) e^{-\varkappa(C_s)|\omega|} , \tag{35^{I}}$$

where the quantities $\varkappa(C_A)$, $\varkappa(C_s)$ satisfy the condition

$$\varkappa(C_A) + \varkappa(C_s) = \varkappa(C_A, C_s) . \tag{35^{II}}$$

It is reasonable to assume that in the calculation of functions such as $\mathcal{J}_3(y)$ expressions of the type given by Eq. (35^{I}) are sufficiently reliable approximations for $\mathcal{J}_0^*(\lambda, x)$. Consequently we shall assume that

$$\mathcal{J}_0^*(\lambda, \omega) = \pi K(\lambda) e^{-\varkappa(\lambda)|\omega|} , \quad \text{where} \quad \lambda = C_{1A}, C_{2A}, C_s . \tag{36}$$

To determine the quantities $\varkappa(C_s)$, $\varkappa(C_{1A})$, $\varkappa(C_{2A})$ the final expression for $\mathcal{J}_3$ must satisfy the conditions of Eqs. (31) and (32).

On combining Eqs. (36) and (34) we obtain an expression for the quantity in which we are interested which, after calculation of the simple integrals in it, assumes the form

$$\mathcal{J}_3(C_{1A}, C_{2A}, C_s; \Delta, y) = \pi K(C_{1A}) K(C_{2A}) K(C_s) \times$$
$$\times \frac{[\varkappa(C_{1A}) + \varkappa(C_{2A}) + \varkappa(C_s)][l][m][n] + y^2 \varkappa(C_s)[\varkappa(C_{1A}) + \varkappa(C_{2A})]^2 + \Delta^2 \varkappa(C_{2A})[\varkappa(C_{1A}) + \varkappa(C_s)] - 2\Delta y \varkappa(C_{2A}) \varkappa(C_s)}{\{[\varkappa(C_{2A}) + \varkappa(C_{1A})]^2 + \Delta^2\}\{[\varkappa(C_{1A}) + \varkappa(C_s)]^2 + y^2\}\{[\varkappa(C_s) + \varkappa(C_{2A})]^2 + (y - \Delta)^2\}} ,$$

where

$$[l][m][n] = [\varkappa(C_{1A}) + \varkappa(C_{2A})][\varkappa(C_{1A}) + \varkappa(C_s)][\varkappa(C_s) + \varkappa(C_{2A})] . \tag{37}$$

It can easily be proved that Eq. (37) satisfies the symmetry requirements (3) without any restrictions on the quantity $\varkappa(\lambda)$. The condition (32^{I}) is satisfied exactly if it is assumed that

$$\varkappa(C_{1A}) + \varkappa(C_{2A}) = \varkappa(C_{2A}, C_{2A}) . \tag{37^{I}}$$

Applying to Eq. (37) the symmetry rules of Eq. (31) we obtain the relationship

$$\varkappa(C_{1A})+\varkappa(C_s)=\varkappa(C_{1A},C_s). \tag{37^{II}}$$

To determine the parameters $\chi(\lambda)$ the relationships (37^{I}) and (37^{II}) must be supplemented by another equation which, strictly speaking, should have such a form that the condition (32) is satisfied, but in view of Eqs. (37^{I}) and (37^{II}) it is more natural to assume that

$$\varkappa(C_{2A})+\varkappa(C_s)=\varkappa(C_{2A},C_s). \tag{37^{III}}$$

In fact in this case the "laws of addition" of the widths of the emission spectrum and the widths of the absorption line components are identical. Finally, from Eqs. (37^{I}), (37^{II}), (37^{III}) we obtain

$$\varkappa(\lambda)=\frac{\varkappa(\lambda,\mu)+\varkappa(\lambda,\nu)-\varkappa(\mu,\nu)}{2}, \tag{38}$$

where λ, μ, ν assume the values C_{1A}, C_{2A}, C_s. Let us determine to what extent the formulas (37) and (38) satisfy the condition (34). By simple transformations we obtain from Eqs. (37) and (38)

$$J_3(C_{1A},C_{2A},C_s;0,y)=$$
$$=\pi K(C_s)\left\{\frac{K(C_{1A})\varkappa(C_{1A},C_s)}{[\varkappa(C_{1A},C_s)]^2+y^2}+\frac{K(C_{2A})\varkappa(C_{2A},C_s)}{[\varkappa(C_{2A},C_s)]^2+y^2}-\frac{K(C_{1A}+C_{2A})\varkappa(C_{1A}+C_{2A},C_s)}{[\varkappa(C_{1A}+C_{2A},C_s)]^2+y^2}P(y)\right\}. \tag{39}$$

Eq. (39) is identical to Eq. (32) when $P(y)\equiv 1$. Without writing out the cumbersome expressions for $P(y)$ we note that, in general, $P(y)\not\equiv 1$; the average value of $P(y)$ over the observed distribution $\varepsilon_2(y)$ is exactly equal to 1. Actual calculations showed that in the range of variation of y which is of practical interest the value of $P(y)$ does not differ by more than a few per cent from 1. A considerable departure of $P(y)$

from 1 only occurs in the "wings" of the distribution $\varepsilon_2(y)$ [$P(y) < 1$]. However, this departure is of little importance. Consequently we obtain

$$\mathcal{J}_3(C_{1A}, C_{2A}, C_s; \Delta, y) = \frac{\pi}{2} K(C_{1A}) K(C_{2A}) K(C_s) \times$$

$$\times \frac{\varkappa(C_{1A}, C_{2A})\varkappa(C_{2A}, C_s)\varkappa(C_{1A}, C_s)[k] + \Delta^2 \varkappa(C_{1A}, C_s)[l] + y^2 \varkappa(C_{2A}, C_{1A})[m] - \Delta y [m][l]}{[\varkappa^2(C_{1A}, C_{2A}) + \Delta^2][\varkappa^2(C_{2A}, C_s) + y^2][\varkappa^2(C_{2A}, C_s) + (y - \Delta)^2]}, \qquad (40)$$

where

$$[k] = [\varkappa(C_{1A}, C_{2A}) + \varkappa(C_{2A}, C_s) + \varkappa(C_s, C_{1A})]$$
$$[l] = [\varkappa(C_{1A}, C_{2A}) + \varkappa(C_{2A}, C_s) - \varkappa(C_s, C_{1A})]$$
$$[m] = [\varkappa(C_{1A}, C_s) + \varkappa(C_{2A}, C_s) - \varkappa(C_{1A}, C_{2A})]$$

A general expression is obtained by combining Eqs. (28^{I}), (28^{II}) and (40). In view of the complexity of the expressions we shall limit our discussion to the case of components of the same intensity ($C_{1A} = C_{2A} = C_A$). With this limitation the observed distribution becomes

$$\varepsilon_2(C_A, C_A, C_s; \Delta, y) = \alpha f K(C_A)\varkappa(C_A, C_s)\left\{\frac{1}{\varkappa^2(C_A, C_s) + y^2} + \frac{1}{\varkappa^2(C_A, C_s) + (y - \Delta)^2} - \right.$$

$$\left. - \frac{K(C_A)\varkappa(C_A, C_A)}{\varkappa^2(C_A, C_A) + \Delta^2}\left(\frac{\left[1 + \frac{\varkappa(C_A, C_A)}{2\varkappa(C_A, C_s)}\right]\varkappa^2(C_A, C_s) + \frac{\Delta^2}{2} + y(y - \Delta)\left(1 - \frac{\varkappa(C_A, C_A)}{2\varkappa(C_A, C_s)}\right)}{[\varkappa^2(C_A, C_s) + y^2][\varkappa^2(C_A, C_s) + (y - \Delta)^2]}\right)\right\} \qquad (41)$$

On the basis of Eq. (41) we can calculate the characteristics of the observed spectrum. We shall first find the positions of the absorption extrema. The general condition for an extremum $\partial\varepsilon_2/\partial y = 0$ leads to an algebraic equation of fifth degree. Solving this equation becomes considerably easier if we use the symmetry properties of the function $\varepsilon_2(y)$. The calculations are given in Appendix III; here we shall merely give the general results. Let us write first the formulas for the quantities C and D which are important in the later treatment.

$$C=\frac{\varkappa^2(C_A,C_s)\left\{\varkappa^2(C_A,C_s)\left[1-\frac{K(C_A)\varkappa(C_A,C_s)}{2[\varkappa^2(C_A,C_A)+\Delta^2]}\left(1-\frac{3}{2}\frac{\varkappa(C_A,C_A)}{\varkappa(C_A,C_s)}\right)\right]-\frac{\Delta^2}{2}\frac{K(C_A)\varkappa(C_A,C_A)}{\varkappa^2(C_A,C_A)+\Delta^2}\cdot\frac{\varkappa(C_A,C_A)}{2\varkappa(C_A,C_s)}\right\}}{1-\frac{K(C_A)\varkappa(C_A,C_A)}{\varkappa^2(C_A,C_A)+\Delta^2}\left[\frac{1}{2}-\frac{\varkappa(C_A,C_A)}{4\varkappa(C_A,C_s)}\right]},$$

$$D=\frac{1}{1-\frac{K(C_A)\varkappa(C_A,C_A)}{\varkappa^2(C_A,C_A)+\Delta^2}\left[\frac{1}{2}-\frac{\varkappa(C_A,C_A)}{4\varkappa(C_A,C_s)}\right]}\times \tag{42}$$

$$\times\left\{\varkappa^2(C_A,C_s)\left[2-\frac{K(C_A)\varkappa(C_A,C_A)}{\varkappa^2(C_A,C_A)+\Delta^2}\left(1+\frac{\varkappa(C_A,C_A)}{2\varkappa(C_A,C_s)}\right)\right]+\Delta^2\left[1-\frac{K(C_A)\varkappa(C_A,C_A)}{\varkappa^2(C_A,C_A)+\Delta^2}\right]\right\}.$$

The results of interest to us can be expressed as follows:

1. If the inequality $\Delta^2<2[D-\sqrt{D^2-4C}]$ is satisfied, then the observed distribution is in the form of a single line with a maximum $y = \Delta/2$; there is no splitting. The value of the maximum absorption is given by the value of $\varepsilon_2(C_A, C_A, C_s; \Delta, y)$ when $y = \Delta/2$. The dependence of $\varepsilon_2(\Delta/2)$ on C_A is given in Fig. 3. The half-width of the absorption line is found from

$$\varepsilon_2\left(C_A,C_A,C_s;\Delta,\frac{\Delta}{2}+\varkappa_\Delta\right)=\frac{1}{2}\varepsilon_2\left(C_A,C_A,C_s;\Delta,\frac{\Delta}{2}\right).$$

The function $\varkappa_\Delta(C_A)$ is given in Fig. 4.

2. $\Delta^2 = 2[D - \sqrt{D^2 - 4C}]$; in this case the first three products in $\varepsilon_2(y)$ become 0 when $y = \Delta/2$ and the absorption spectrum represents a "compact" maximum.

3. When $\Delta^2 > 2[D - \sqrt{D^2 - 4C}]$ the distribution $\varepsilon_2(y)$ has three extremal points (splitting of the true spectrum): these are a minimum at $y = \Delta/2$ and two symmetrically located maxima at the points

$$y_{1,2}=\frac{\Delta}{2}\pm\frac{1}{2}\sqrt{\Delta^2-2[D-\sqrt{D^2-4C}]}. \tag{43}$$

The separation between the maximum points of Δ_{obs} are given by

$$\left(\frac{\Delta_{obs}}{\Delta}\right)^2=1-\frac{2}{\Delta^2}[D-\sqrt{D^2-4C}]. \tag{44}$$

Hence Δ_{obs} is always smaller than Δ because $C > 0$. The difference between the observed and true splitting is very important in the cases of practical interest, as shown in Fig. 5. For example, in the case of infinitely thin sources and absorbers the true and observed splittings are equal only for $\Delta > 6$.

The values of the maximum and minimum absorption are given by the relationships

$$\mathcal{E}_{2\,max} = \mathcal{E}_2(y_1) = \mathcal{E}_2(y_2) \quad ; \quad \mathcal{E}_{min} = \mathcal{E}_2\left(\frac{\Delta}{2}\right). \tag{45}$$

We note that when $\Delta = 0$ the expression for the maximum absorption does not reduce to the corresponding formula for a single line; it can easily be seen that this is due to the fact that the selected approximation for the Fourier transformation $J_0(\lambda, x)$ gives incorrectly the "high-frequency" part of the spectrum of the function, ie. it is found (Appendix I) that the "high" frequencies are somewhat overestimated. From the practical point of view the error involved is of little importance since the difference between Eqs. (23) and (45) for $2C_A \leqslant 10$, $C_s \leqslant 10$ does not exceed 3%.

In conclusion, we note the following point. Assuming that an analysis of the experimental results gives the quantities f', Δ, Γ, αf, then as a check it is useful to measure the absorption away from the middle of the spectrum. This absorption should be

$$\mathcal{E}_{check} = \mathcal{E}_2(C_A, C_A, C_s; \Delta, 0) = \frac{\alpha f}{K(C_s)}\Big[K(C_A) + K(C_s) - K(C_A + C_s) + \frac{K(C_A+C_s)\cdot K(C_A)\cdot \mathcal{H}(C_A+C_s, C_A)}{\mathcal{H}^2(C_A+C_s, C_A) + \Delta^2} - \frac{K^2(C_A)\cdot \mathcal{H}(C_A, C_A)}{\mathcal{H}^2(C_A, C_A) + \Delta^2}$$

The authors take this opportunity of thanking Professor V. S. Shpinel' for his constant interest and all possible support in this work.

APPENDIX I

It has been shown above that for single lines the shape of the experimental spectrum is close to the dispersion curve. Let us seek the reason for this. We shall start from the formula (6)

$$\varepsilon(y) = \frac{1}{\pi}\int_{-\infty}^{+\infty}\left[1 - \exp\left(-\frac{C_A}{1+x^2}\right)\right]\frac{dx}{1+(x+y)^2} \qquad (\alpha f = 1). \tag{AI-1}$$

On the basis of the above-mentioned theorem on convolution-type functions we can write

$$\varepsilon^*(w) = J_0^*(C_A, w)\cdot e^{-|w|} \tag{AI-2}$$

where $\varepsilon^*(w)$ is the Fourier transformation of $\varepsilon(y)$: from (AI-2) it follows that the statement that $\varepsilon(y)$ is close to the dispersion function $\left(\varepsilon^*(w) \simeq \pi K(C_A)\, e^{-\varkappa(C_A,0)|w|}\right)$ is equivalent to

$$J_0^*(C_A, w) \simeq \pi K(C_A) \exp\left[-|w|\left(\alpha(C_A,0) - 1\right)\right]$$ 3)

$$J_0(C_A, x) \simeq \frac{K(C_A)\left[\varkappa(C_A,0) - 1\right]}{\left[\varkappa(C_A,0) - 1\right]^2 + x^2} \tag{AI-3}$$ 3)

The validity of the second of the above two equations can be checked easily by direct comparison with the corresponding functions. By way of example Fig. 6 gives $J_0(5, x)$ and the dispersion curve $5.41/(5.175 + x^2)$ which approximates to it according to Eq. (AI-3). The difference between the two curves is small: in the region $0 \leqslant x \leqslant 8$ the difference does not exceed 5%, and at higher values of x the error reaches its maximum value of 8%. This gives us an idea of the precision of the function $\varepsilon(y)$. In fact, let $J_0(C_A, x)$ be convoluted with

---

3) In view of their relative simplicity, approximations of this type are widely used in studies of convolution-type expressions /7/.

some quite general function, which we shall denote by $\varphi(x)$; the convolution of these functions we shall denote by $\varepsilon_\varphi(y)$. It is easily seen that if $\varphi(x) = const\ (\varepsilon_\varphi = \pi K(C_A) const)$, then on replacing $J_0(C_A, x)$ by the dispersion curve which approximates to it, the result of convolution is unaffected in the second limiting case when $\varphi(x)$ is of delta-type $[\varphi(x) = \delta(x)]$ and the result of the convolution is identical with the function $J_0(C_A, x)$; the error in $\varepsilon_\varphi(x)$ is then equal to the error in the approximation to $J_0(C_A, x)$. In the intermediate cases when $\varphi(x)$ is represented by some finite width $\Delta\varphi$ the error in the convolution $\varepsilon_\varphi(x)$ lies between these limits and, in the first approximation, is determined by the ratio of the widths of the convoluted functions $\varphi_0(x)$ and $J_0(C_A, x)$. In our case $\varphi(x) = 1/[\pi(1 + x^2)]$ and the ratio of the widths of $\varphi(x)$ and $J_0(C_A, x)$ is $1/[\varkappa(C_A, 0) - 1]$. For qualitative estimates we may assume $\sigma_\varepsilon = \sigma_{J_0}/[1 + (\Delta_\varphi/\Delta_{J_0})A]$, where σ_ε is the relative "average" error in the convolution of the functions $\varphi(x)$ and $J_0(C_A, x)$; σ_{J_0} is the average error permissible when $J_0(C_A, x)$ is replaced by the dispersion distribution; A is a constant for a fixed distribution $\varphi(x)$. Δ_φ and Δ_{J_0} denote the widths of the corresponding distributions.

To find A it is sufficient to compare σ_ε and σ_{J_0} in some particular case. Using a method which will be referred to later, such a comparison was carried out for $C_A = 1$; the quantity to be found was ≈ 3.5. Thus we see that the error in the $\varepsilon(y)$ distribution is considerably smaller than the error permissible in the approximation to $J_0(C_A, x)$ by the dispersion curve. This is clearly illustrated in Table 1 for $C_A = 1$ where the notation $\eta_{\mathcal{I}} = \frac{\mathcal{I}_{o\,guen} - \mathcal{I}_o(1,y)}{\mathcal{I}_o(1,y)}$, $\eta_{\mathcal{E}} = \frac{\mathcal{E}_{guen} - \mathcal{E}(y)}{\mathcal{E}(y)}$ is used. Let us check now the first of the relationships in Eq. (AI-3). In view of the parity of $\mathcal{I}_o(C_A, x)$ we can write $\mathcal{I}_o^*(C_A, w) = 2\int_0^\infty \left[1 - \exp\left(-\frac{C_A}{1+x^2}\right)\right] \cos wx \, dx$.

Expanding $\mathcal{I}_o(C_A, x)$ as a series in powers of C_A we obtain

$$\mathcal{I}_o(C_A, x) = C_A \sum_{n=0}^{\infty} (-1)^n \frac{C_A^n}{(n+1)!} \frac{1}{(1+x^2)^{n+1}} \, . \tag{AI-5}$$

Hence $\mathcal{I}_o^*(C_A, w)$ is written as

$$\mathcal{I}_o^*(C_A, w) = 2C_A \sum_{n=0}^{\infty} (-1)^n \frac{C_A^n}{(n+1)!} \int_0^\infty \frac{\cos wx}{(1+x^2)^{n+1}} dx \, . \tag{AI-6}$$

According to /5/ the integrals in (AI-6) are given by

$$\int_0^\infty \frac{\cos wx}{(1+x^2)^{n+1}} dx = \sqrt{\pi} \left|\frac{w}{2}\right|^{n+\frac{1}{2}} \frac{1}{n!} K_{n+\frac{1}{2}}(|w|), \tag{AI-7}$$

where $K_{n+(1/2)}(\lambda)$ is a Hankel function of half-integer order with an imaginary argument. Combining (AI-6) and (AI-7) we obtain

$$\mathcal{I}_o^*(C_A, w) = C_A \sqrt{2\pi|w|} (-1)^n \left[\frac{C_A |w|}{2}\right]^n \frac{1}{(n+1).\, n!} K_{n+\frac{1}{2}}(|w|) \, . \tag{AI-8}$$

For a half-integer index the K-function reduces to the elementary functions

$$K_{n+\frac{1}{2}}(\lambda) = \left(\frac{\pi}{2\lambda}\right)^{1/2} e^{-\lambda} \sum_{r=0}^{\infty} \frac{(n+r)!}{r!\,(n-r)!\,(2\lambda)^r} \, . \tag{AI-9}$$

The relationships (AI-8) and (AI-9) allow us to construct the exact spectrum of the function $\mathcal{I}_o^*(C_A, w)$. In practical calculations it is useful to employ the recurrent relationships between the K-functions:

$$K_{p+1}(\lambda) = \frac{2p}{\lambda} K(\lambda) + K_{p-1}(\lambda) \, .$$

We compared the exact spectrum of the function $\mathcal{I}_o^*(1, w)$ and the spectrum of the approximating dispersion curve $\mathcal{I}_{disp}^*(1, w)$. Table 2 gives the functions $\mathcal{I}_o^*(1, w)$ and $\mathcal{I}_{disp}^*(1, w)$. The series in Eq. (AI-8) allows us to determine the constant in (AI-4).

We shall represent $\mathcal{J}_0^*(C_A, w)$ in the following way

$$\mathcal{J}^*(C_A, w) = \pi e^{-|w|} \sum_{n=0}^{\infty} f_n(C_A)|w|^n . \tag{AI-13}$$

The form of the function $f_n(C_A)$ is established by comparing the series (AI-8) and (AI-13), taking into account Eq. (AI-9). Combining Eqs. (AI-2) and (AI-13) and using the formula /5/

$$\int_0^{\infty} e^{-qx} x^{p-1} dx = q^{-p} \Gamma(p) , \tag{AI-14}$$

we obtain the following expression for $\varepsilon(y)$:

$$\begin{aligned} \varepsilon(y) &= \frac{1}{2} \sum_{n=0}^{\infty} f_n(C_A) \Gamma(n+1) \left[\frac{1}{(2-iy)^{n+1}} + \frac{1}{(2+iy)^{n+1}} \right] = \\ &= \frac{2K(C_A)}{[4+y^2]} + \frac{[4+y^2] f_1(C_A)}{[4+y^2]^2} + \frac{2f_2(C_A)[8-6y^2]}{[4+y^2]^3} + \\ &+ \frac{6 f_3(C_A)[16-12y^2+y^4]}{[4+y^2]^4} + \frac{24 f_4(C_A)[32-80y^2+10y^4]}{[4+y^2]^5} + \dots \end{aligned} \tag{AI-15}$$

Eq. (AI-15) is used to determine the constants A in (AI-4). Summing up, we can conclude that in the region of variation of C_A which is of practical interest the observed distribution is of dispersion type.

APPENDIX II

The condition for an extremum $\partial \varepsilon_2(y)/\partial y = 0$ leads to the following algebraic equation

$$\begin{aligned} &(2-B)y^5 + 5\Delta\left(\frac{B}{2}-1\right)y^4 + 2\left[2\varkappa^4(C_A,C_S) + 3\Delta^3 - B\Delta^2 - A\right]y^3 + \\ &+ \Delta\left[3A + \frac{B}{2}\Delta^2 - 4\Delta^2 - 6\varkappa^2(C_A,C_S)\right]y^2 + \left[2\varkappa^4(C_A,C_S) + 2\varkappa^2(C_A,C_S)\Delta^2 + \Delta^4 + \right. \\ &\left. + B\varkappa^4(C_A,C_S) + B\Delta^2\varkappa^2(C_A,C_S) - 2A\varkappa^2(C_A,C_S) - \Delta^2 A\right]y + \\ &+ \varkappa^2(C_A,C_S)\left[A - \frac{B\varkappa^4(C_A,C_S)}{2} - \frac{B\Delta^2}{2} - \varkappa^2(C_A,C_S)\right] = 0 . \end{aligned} \tag{AII-1}$$

Here

$$\begin{aligned} A &= \frac{K(C_A)\varkappa(C_A,C_S)}{\varkappa^4(C_A,C_S)+\Delta^2} \left\{ \left[1 + \frac{\varkappa(C_A,C_A)}{2\varkappa(C_A,C_S)}\right] \varkappa^2(C_A,C_S) + \frac{\Delta^2}{2} \right\} \\ B &= \frac{K(C_A)\varkappa(C_A,C_A)}{\varkappa^2(C_A,C_A)+\Delta^2} \left\{ 1 - \frac{\varkappa(C_A,C_A)}{2\varkappa(C_A,C_S)} \right\} . \end{aligned} \tag{AII-2}$$

Due to the symmetry properties of $\varepsilon_2(y)$ the roots of (AII-1) obey the following equalities:

$$y_1 = \frac{\Delta}{2}, \quad y_2 + y_3 = \Delta \quad ; \quad y_4 + y_5 = \Delta . \tag{AII-3}$$

Using (AII-3) and the general relationships between the roots of algebraic equations /8/ we obtain:

$$y_2(\Delta - y_2) - y_3(\Delta - y_3) = C \quad ; \quad y_1(\Delta - y_1) + y_3(\Delta - y_3) = D , \tag{AII-4}$$

where C and D are given by Eq. (42). Hence we can easily find that the real roots of (AII-1) have the form

$$y_{1,2} = \frac{\Delta}{2} \pm \frac{1}{2}\sqrt{\Delta^2 - 2\left[D - \sqrt{D^2 - 4C}\right]} \tag{AII-5}$$

APPENDIX III

In conclusion, something should be said about a simultaneous allowance for the resonance and electron self-absorption in the source. In this case all the conclusions derived for the resonant self-absorption in the source still apply, with the difference that in the appropriate formulas the function $K(\alpha)$ should be replaced by a function $L(\alpha, \beta, \gamma)$ defined as

$$L(\alpha, \beta, \gamma) = \frac{1}{\pi}\int_{-\infty}^{+\infty} \frac{1 - \exp - \left(\frac{\alpha}{1+x^2} + \beta\right)}{1 + \gamma(1 + x^2)}\, dx . \tag{AIII-1}$$

For the integral in (AIII-1) we obtain

$$L(\alpha, \beta, \gamma) = \frac{1}{\sqrt{\gamma^2 + \gamma}}\left\{(1 - e^{-\beta}) + \frac{e^{-\beta} K(\alpha)}{\mathcal{X}(x) - 1 + \sqrt{\frac{1}{\gamma} + 1}}\right\} . \tag{AIII-2}$$

The parameters of the spectrum then become

$$S_{\text{exp}} = \alpha f \pi L(C_A, 0, 0) , \tag{AIII-3}$$

$$\mathcal{E}(0) = \frac{\alpha f}{L\left(C_s, C_e, \frac{C_e}{C_s}\right)}\left[L\left(C_A, 0, \frac{C_e}{C_s}\right) + L\left(C_s, C_e, \frac{C_e}{C_s}\right) - L\left(C_A + C_s, C_e, \frac{C_e}{C_s}\right)\right],$$

$$\mathcal{X}(C_A, C_s, C_e) = \frac{L(C_A, 0, 0)\, L\left(C_s, C_e, \frac{C_e}{C_s}\right)}{L\left(C_A, 0, \frac{C_e}{C_s}\right) + L\left(C_s, C_e, \frac{C_e}{C_s}\right) - L\left(C_A + C_s, C_e, \frac{C_e}{C_s}\right)} , \tag{AIII-4}$$

(AIII-5)

where $C_e = \mu_e d$, μ_e is the electron absorption coefficient, and d is the source thickness. It follows however that the influence of electron absorption on the parameters of the observed spectrum is extremely small.

Institute of Nuclear Physics, Moscow State University

Table 1

y	$\eta_{\gamma_0}(\pm\%)$	$\eta_E(\pm\%)$	y	$\eta_{\gamma_0}(\pm\%)$	$\eta_E(\pm\%)$	y	$\eta_{\gamma_0}(\pm\%)$	$\eta_E(\pm\%)$	y	$\eta_{\gamma_0}(\pm\%)$	$\eta_E(\pm\%)$
0	+ 0,89	0.00	3	- 0.07	- 0,II	6	+ 0,50	+ 0,I6	9	+ 0,57	+. 0,27
I	- 0,82	0.00	4	+ 0.23	0,00	7	+ 0,52	+ 0,20	I0	+ 0.57	+ 0,27
2	- 0,50	-0,20	5	+ 0,34	+ 0,09	8	+ 0,55	+ 0.24	II	+ 0,57	+ 0,27

Table 2

w	$J_0^*(1,w)$	$J_{gac}^*(1,w)$	w	$J_0^*(1,w)$	$J_{gac}^*(1,w)$	w	$J_0^*(1,w)$	$J_{gac}^*(1,w)$	w	$J_0^*(1,w)$	$J_{gac}^*(1,w)$
0	2,5I6	2.5I6	I.0	0,7I8	0,7I3	2,5	0.I02	0.I08	4.0	0.0I4	0,0I6
0,2	I.965	I.959	I.5	0,376	0.380	3.0	0.052	0,060	4.5	0.006	0.008
0.5	I.36I	I.340	2.0	0,20I	0,202	3.5	0.027	0.030	5,0	0.003	0.005

REFERENCES

1. S. Margulies, J. R. Ehrman, Nucl. Instr. and Methods 12, 131-7 (1961).
2. D. A. Shirley, M. Kaplan, P. Axel, Phys. Rev. 123, 816 (1961).
3. N. N. Delyagin, V. S. Shpinel', V. A. Bryukhanov, ZhETF 41, 1347 (1961).
4. R. F. Mössbauer, W. H. Wiedemann, Z. Physik 159, 33 (1960).
5. M. Born, Optik (Berlin, 1933) /Russian translation/ (ONTI, 1937).
6. K. M. Ryzhik, Gradshtein, Table of Integrals, Sums, Series and Products /in Russian/ (Gostekhizdat, M.-L., 1956), 3rd edition.
7. E. Titimarsh, Introduction to the Theory of Fourier Integrals /in Russian/ (Gostekhizdat, M.-L., 1948).
8. S. G. Rautian, UFN 66, 475 (1958).
9. E. S. Lyapin, Course of Higher Algebra /in Russian/ (Uchpedgiz, 1953).

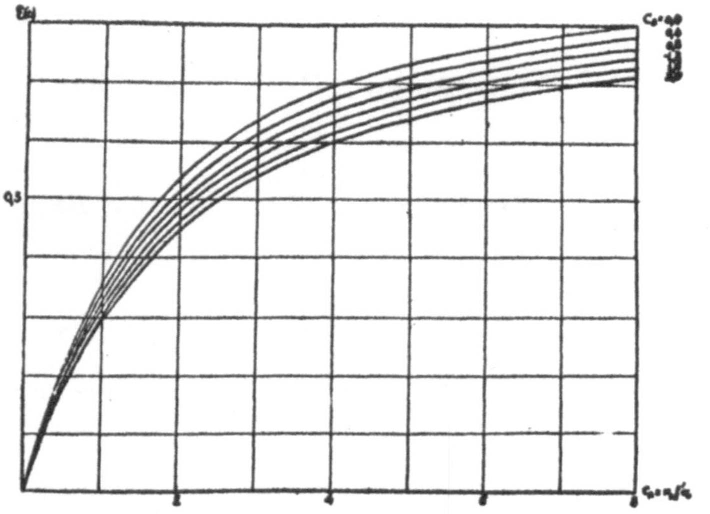

Fig. 1. Dependence of the maximum absorption $\varepsilon(0)$ on the parameters C_A and C_s.

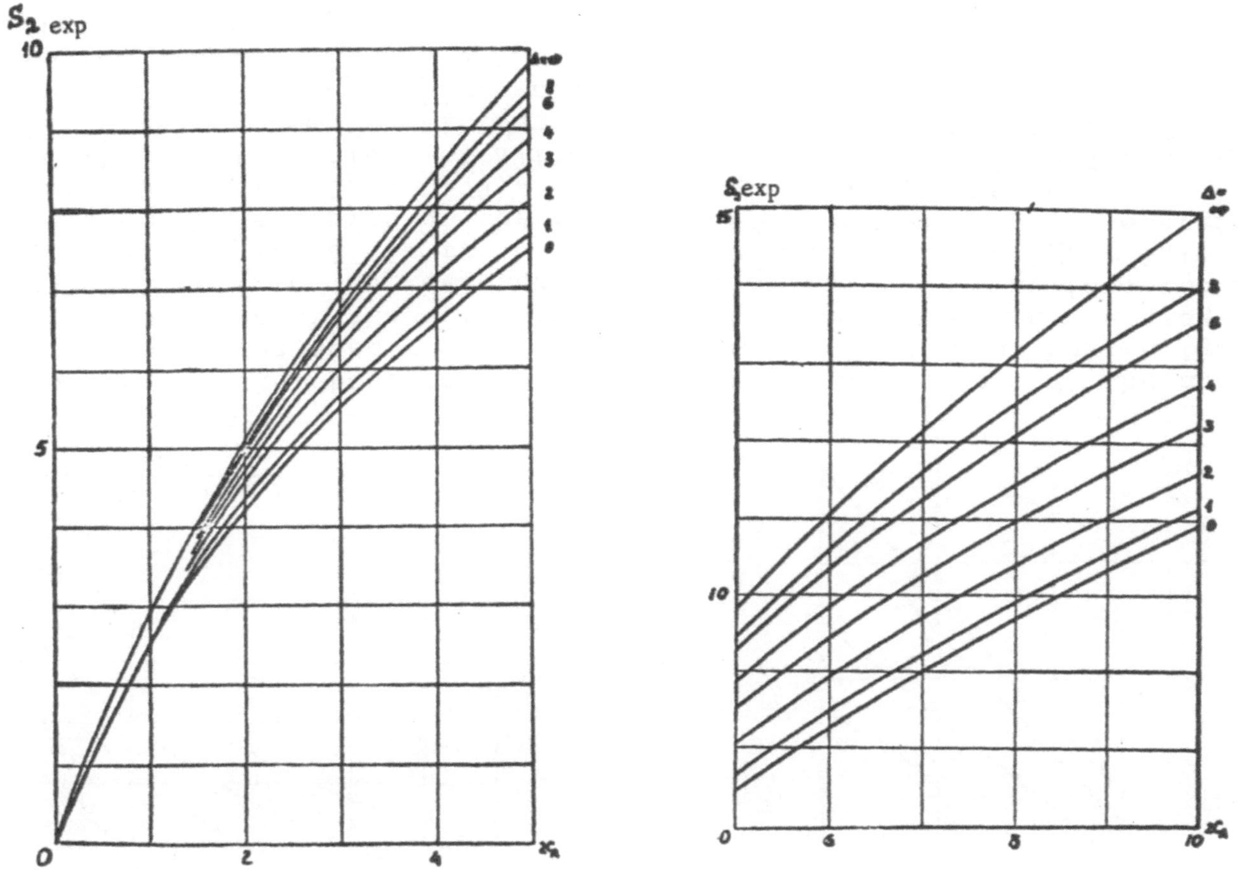

Fig. 2. Dependence of the integrated absorption on the parameter C_A in the case of the splitting of an absorption line into two identical components $(C_{1A} = C_{2A} = C_A)$. Δ is the difference of the energies of the components in units of $\Gamma/2$. Here $S_{exp} = \int_{-\infty}^{\infty} \varepsilon(y)dy$, where $y = (v/c)(E_0/\frac{1}{2}\Gamma)$.

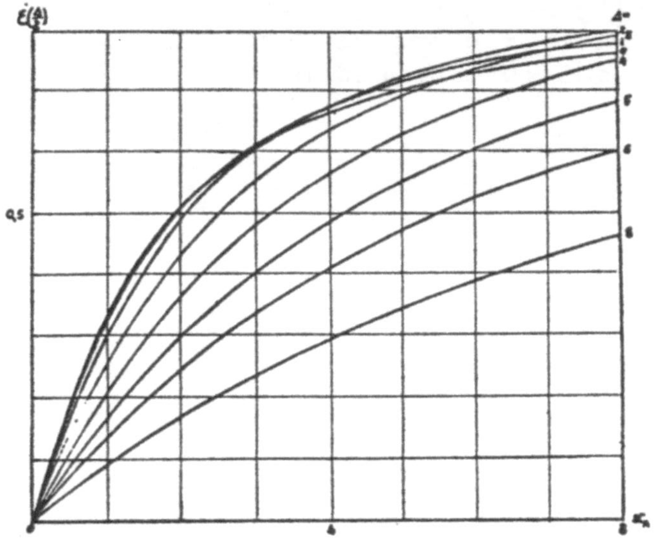

Fig.3. Dependence of the absorption at the mid-point of the observed spectrum on the parameter C_A in the case of the splitting of an absorption line into two identical components ($C_{1A} = C_{2A} = C_A$). Δ is the difference between the energies of the components in units of $\Gamma/2$; ($C_s = 0, 4$). At high values of C_A and low values of Δ the quantity $\varepsilon(\Delta/2)$ is a non-monotonic function of Δ. This property appears at high values of the absorption and is related to the presence of a region where the components overlap.

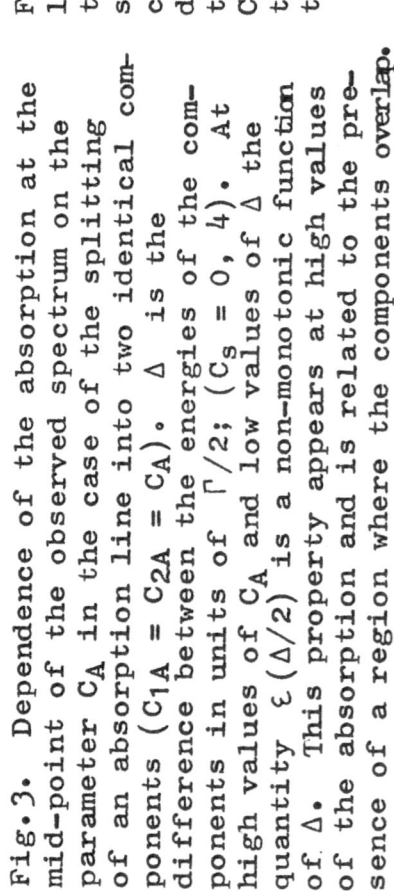

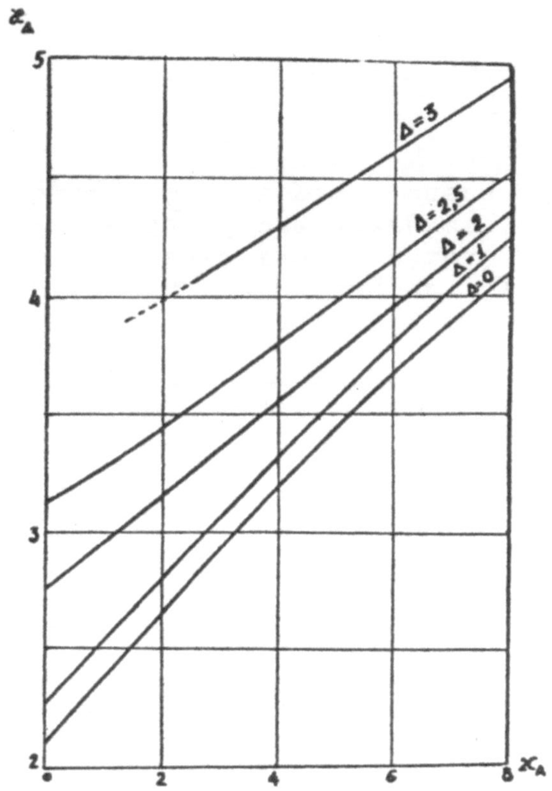

Fig.4. Dependence of the absorption line half-width at half-height, $\varkappa_\Delta$, on the parameter C_A in the case of the splitting of a line into two identical components ($C_{1A} = C_{2A} = C_A$); Δ is the difference between the energies of the components in units of $\Gamma/2$; $C_s = 0, 4$. The dashed line represents the region of variation of C_A where the observed spectrum is split.

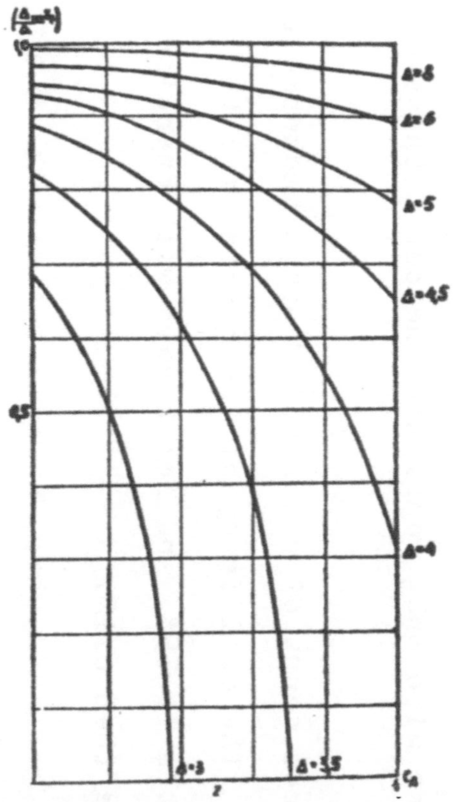

Fig.5. Dependence of the ratio (Δ_{obs}/Δ) on the parameter C_A in the case of the splitting of an absorption line into two identical components $(C_{1A} = C_{2A} = C_A)$; Δ_{obs} is the distance between the absorption maxima of the observed spectrum in units of $\Gamma/2$; Δ is the difference between the energies of the components in units of $\Gamma/2$ $(C_s = 0, 4)$.

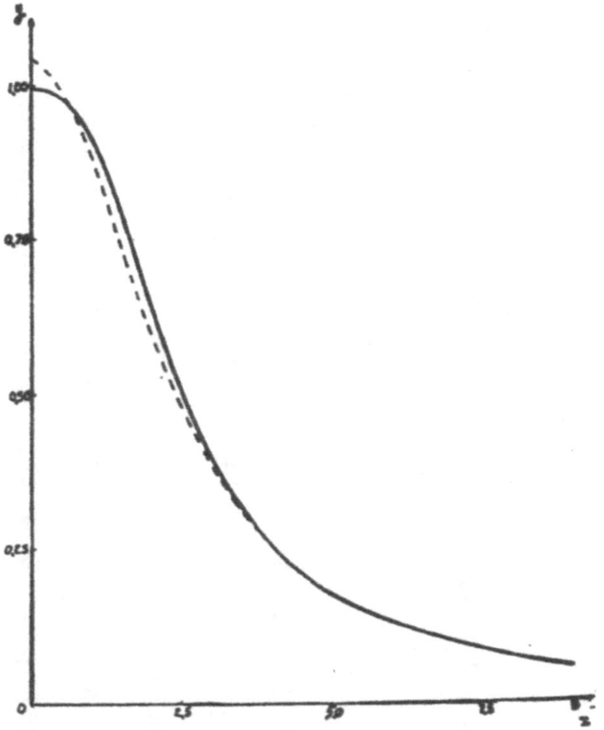

Fig.6. Comparison of the function (continuous curve) with the dispersion function which approximates to it, (dashed curve). The figure represents the case $C_A = 5$.

A COUNTER WITH SELECTIVE EFFICIENCY FOR RECORDING THE RECOILLESS γ-RAYS OF Sn^{119}

K. P. Mitrofanov, N. V. Illarionova, V. S. Shpinel'

A detector which has selective efficiency for the 23.8-keV γ-rays emitted without recoil from Sn^{119} is proposed and tested. The principle of its operation is based on the recording of internal-conversion electrons emitted by resonance-excited nuclei in a tin compound which has been placed within a Geiger counter.

A calculation is made of the shape of an absorption line measured by a "resonance" detector. It is shown that its width is less than that obtained by the ordinary method of recording.

An experimental test made with a flat resonance counter whose inner surface was coated with SnO_2 confirmed the correctness of the calculation.

The maximum measured count of resonance γ-rays amounted to 250% above background. The efficiency of the counter for the recoilless γ-rays of tin is 15%, and for x-rays with energies of 25 keV it is no more than 0.1%.

1. INTRODUCTION

The Mössbauer effect is usually observed by a transmission method, in which the attenuation of a γ-ray beam by absorbers containing resonance-absorbing nuclei is determined. A high counting rate is obtained by this method, but the magnitude of the effect is small. (By the effect we mean the change in count rate due to resonance absorption relative to the count rate without resonance absorption.) Nuclei which have been excited as a result of resonance absorption of γ-rays emit γ-quanta or internal-conversion electrons and x-rays, which also may be used to observe the Mössbauer effect. In experiments in which secondary radiation is recorded it is possible to obtain an effect of considerably greater magnitude. The weakness of this method is the small counting rate.

These experiments, as a rule, are done with γ-rays or x-rays since their greater penetrating power permits the intensity of the scattered radiation to be increased by using absorbers of greater thickness.

The use of conversion electrons for recording resonance absorption is limited by the small value of their range. However, in those cases in which the cross section for resonance capture of γ-rays is so large that an absorber which is thin with respect to conversion electrons is thick with respect to recoilless γ-rays, the use of electron recording can prove to be more efficient /1, 2/.

The 23.8-keV γ-transition in Sn^{119} ($\alpha = 6.3$) is one of the most favorable cases for using electron recording. Internal conversion of 23.8-keV γ-rays occurs in the L, M, etc. shells, as a result of which the conversion electrons have energies close to the energy of the γ-transition and ranges comparable with the mean free path of the recoilless γ-rays. (This condition is valid for the chemical compounds of tin for which the probability of resonance absorption of γ-rays is rather large, e.g., for SnO_2 /3/.)

Since the cross section for resonance absorption of γ-rays is considerably greater than the cross section for the photoelectric effect, the electron component of the secondary emission of an SnO_2 absorber which is irradiated by γ-rays from Sn^{119*} consists mainly of conversion electrons. This fact permits a considerable simplification in the technique of recording resonance absorption. As a matter of fact, placing a layer of SnO_2 (the radiator) inside a Geiger counter is sufficient to produce 100% recording of all electrons emitted from the radiator. Such a counter (we shall call it a "resonance counter") will have increased efficiency for recoilless γ-rays.

The use of a resonance counter makes it possible to combine the positive qualities of absorption experiments and experiments which record secondary radiation, i.e., to obtain a large effect with a high counting rate. Observations of resonance absorption are usually made by measuring the dependence of counting rate on the relative velocity of source and absorber (scatterer).

In using a resonance counter we can consider its radiator as an absorber and measure the resonance absorption spectrum by giving the source different values of velocity relative to the counter (Fig. 1a).

When emission and absorption lines are superimposed as a result of the Doppler shift, a counting maximum will be observed. This method of measurement requires that the source or the radiator be made of the substance which is the subject of the investigation, which is not always convenient.

It is much more rational to use a scheme of measurement in which the source and counter are fixed and are in resonance with one another, and a movable absorber of the substance being investigated is placed in the space between them (Fig. 1b). In this case, depending on the speed of the absorber, the number of resonance γ-rays passing through the counter will change, causing a corresponding change in the counting rate. In such a setup the result is an absorption spectrum in which a minimum in the counting level will be observed when an emission line of the source coincides with an absorption line of the absorber. This arrangement is analogous to experiments on the Mössbauer effect which are performed by the absorption method, only the detector is a resonance counter instead of a scintillation spectrometer. The use of a resonance counter in this geometry leads mainly to a considerable decrease in nonresonance contributions to the count. In addition, a detector with a sharp dependence of efficiency on energy can change the shape of a line in comparison with a line measured by an ordinary counter.

2. PROPERTIES OF A RESONANCE DETECTOR IN THE ABSORPTION METHOD OF MEASUREMENT

In the majority of experiments on the Mössbauer effect the shapes and intensities of lines in the absorption spectrum are studied. The simplest absorption spectrum, consisting of a single line, is observed when the emission and absorption lines are single. The intensity and width of a line in an absorption spectrum are known to depend on the number of resonance-absorbing nuclei in the source and in the absorber. The effective thickness of the source and absorber with regard to the isotope in question is characterized by the quantity $C = \sigma_0 f' n$, where σ_0 is the maximum effective cross section for resonance absorption, f' is the probability of resonance absorption of γ-rays, and n is the number of atoms of the resonance-absorbing isotope per cm^2 of surface.

For small thicknesses of source and absorber an analytic expression has been obtained for resonance absorption; it has been shown that a line in the absorption spectrum has a profile which can be described by a Lorentz curve (see, for example, /4/),

$$N(y) \sim 1 - f \frac{C_a}{2} \frac{1}{1+y^2} \tag{1}$$

with a width $\Gamma_0 = 2\Gamma$, where C_a is the thickness of the absorber ($C_a \ll 1$), Γ is the width of the nuclear level, $y = (v/c)(E_0/\Gamma)$ is the relative Doppler shift, and f is the probability of recoilless emission of γ-rays.

If the source or absorber have finite thickness, the absorption line calculation is done by numerical methods.* Calculations of (4) done by an electronic computer for source and absorber thicknesses $0 \leq C_s, C_a \leq 10$ showed that an absorption line retains the form of a Lorentz curve, but its width (Γ') depends on C_s and C_a approximately linearly, while the thicknesses of the source and absorber have about the same effect on the width. The relationship between Γ' and the thickness of the source or absorber can be represented by the expression

$$\Gamma' = 2\Gamma(1 + 0{,}135C)_{0 \leq C \leq 5} \tag{2}$$

---

* There is a method for precise determination of the parameters of absorption lines proposed in /6/ which is based on the measurement of the areas under the appropriate spectra.

Thus, the recoilless component of the γ-spectrum of a thick source can, with good precision, be written as $N(E) = \mathrm{Const}/[(E - E_0)^2 + (b\Gamma/2)^2]$, where $b = \Gamma'/\Gamma$ is the relative broadening of the emission line. We used this expression for a single emission line of a thick source in a comparison of the forms of absorption spectra calculated for measurements with ordinary counters and with resonance counters.

The transmission of recoilless γ-rays in the energy interval from E to E + dE through a resonance absorber which is moving with a velocity v relative to the source is written in the following way:

$$I(E,v) = \frac{\mathrm{Const}\, dE}{(E-E_0)^2+\left(\frac{b\Gamma}{2}\right)^2} \cdot \exp\left[-\frac{\Gamma^2/4 \cdot C_a}{(E-E_0-v/c\,E_0)^2+\Gamma^2/4}\right]$$

If an ordinary detector is used whose efficiency remains constant during a change in γ-ray energy by an amount of the order of Γ, the variation in count of the "resonance" component of the spectrum as a function of the velocity of the absorber, i.e., the absorption curve, is given by the expression

$$N(v) = \int_0^\infty I(E,v)\, dE \qquad (3)$$

In using a resonance detector only that fraction of the γ-rays emitted in the energy range dE will be recorded which, after passing through the moving filter, is absorbed in the radiator, whose thickness we shall take to be equal to C_r. For the absorption curve we obtain the more complex expression

$$N_2(v) = \varepsilon \int_0^\infty I(E,v)\left\{1 - \exp\left[-\frac{\Gamma^2/4\, C_r}{(E-E_0)^2+\Gamma^2/4}\right]\right\} dE \qquad (4)$$

The symbol ε here denotes the average, over the whole volume of the radiator, of the emission probability of a nucleus which has been excited by resonance capture, with a conversion electron striking the effective volume of the counter.

When C_a, $C_r \ll 1$ the integration of (4) leads to a precise expression for the count as a function of v and b, which will not be cited here because it is cumbersome. When b = 1 this expression reduces to the form

$$N_2(y) \sim \left[1 - \frac{C_a(y^2+3)}{4(y^2+1)^2}\right] \tag{5}$$

An absorption line profile calculated for a resonance counter differs somewhat from a Lorentz curve, but is very close to it. The line described by Eq. (5), for example, is well approximated by a Lorentz curve with $\Gamma' = 1.47\Gamma$.

Absorption curves for finite values of C_a and C_r which were plotted from numerical integration of (4) were compared with absorption lines from an ordinary detector which were calculated by numerical integration of (3).

The result of one such series, in which b = 3, $C_a = 1$, and $C_r = 1$, 3, and 10, is presented in Fig. 2 the factor $\mathcal{E}$, which does not lend itself to precise calculation, was taken to be equal to unity in the plotting of these curves. The counting level for high absorber speeds $N_r(\infty)$ corresponds to the maximum possible recording of recoilless γ-rays with a given C_r (broken lines in Fig. 2). (Let us remember that only the resonance component of the γ-spectrum of the source was taken into consideration in the calculations.) For an ordinary detector the count $N(\infty)$ will be equal to the total number of γ-quanta incident on it. In the case of a resonance detector the count $N_r(\infty)$ consists of the part of the total flux which depends on C_r. With increasing C_r the counting level rapidly rises, approaching the limiting value $N(\infty)$. The depth of the counting minimum at $v = 0$, taken as a ratio to the total flux of γ-quanta, $[N_r(\infty) - N_r(0)]/N(\infty) = \eta_o$ characterizes the absolute magnitude of the effect; and the fraction $[N_r(\infty) - N_r(0)]/N_r(\infty) = \eta_r$ determines its relative magnitude.

The values of η_o and η_r determined from the graphs and also the line width Γ' at half depth, for resonance counters of different thicknesses, were compared with the analogous quantities for a nonresonance detector. Curves are presented in Fig. 3 which illustrate the variation of these quantities as a function of C_r for two cases: a thin source ($b = 1$) and a thick source ($b = 3$), with the absorber thickness in both cases given by $C_a = 1$. It is essential that the absolute magnitude of the effect increases sharply with the thickness of the radiator and by $C_r = 2$ has more than 50% of its maximum value. At the same time the relative magnitude of the effect is greater for measurements with a resonance counter than with an ordinary detector, and its decrease with increasing C_r is much more weakly expressed.

As calculations show, lines measured with a resonance counter should be sharper. This can be seen, for example, by comparing the curves of Eq. (1) $\Gamma' = 2\Gamma$ and Eq. (5) $\Gamma' = 1.47\Gamma$. Moreover, the broadening of an emission line due to self absorption in the source, which under ordinary conditions enters as a component in the absorption width, proves to be considerably weaker. For example, with a "thin counter" and an absorber the width of the absorption curve changes only from 1.47Γ to 2Γ as b varies from 1 to ∞.

When a resonance counter is used, the expression for the broadening of an absorption line due to source thickness /Eq. (2)/ is replaced by the expression

$$\Gamma \simeq 2\Gamma/0{,}75 + \delta C_s/, \tag{6}$$

where δ is a quantity less than 0.135 which depends on the thickness of the counter. For $C_r = 1\text{-}3$, $\delta = 0.01\text{-}0.04$.

As the thickness of the counter increases its resonance properties become weaker and weaker, and in the limiting case $C_r \rightarrow \infty$ all recoilless γ-quanta will be absorbed in the radiator independently of their energy in relationship to E. This phenomenon is easily explained by a consideration of Eq. (4), where, as $C_r \rightarrow 0$, the factor which does not depend on v has the form $Const/[(E - E_o)^2 + \Gamma^2/4^2]$, which is equivalent to Eq. (3), but with a sharper dependence on the energy of the emission line. This leads to a corresponding narrowing of the absorption line. With increasing C_r this dependence becomes weaker, and as $C_r \rightarrow \infty$ Eq. (4) goes over into Eq. (3), i.e., the counter loses its resonance properties. Therefore, it is most expedient to use a counter with $C_r \sim 1\text{-}3$, where it appears possible to achieve a considerable increase in the relative effect and a decrease in line width with the absolute magnitude of the effect close to maximum.

3. EXPERIMENTAL TEST OF THE OPERATION OF THE RESONANCE COUNTER

A flat counter has the most convenient construction for operation as a resonance counter since the closely spaced plane surfaces of the cathodes can easily be used for deposition of the radiator. Moreover, if necessary, it is easy to combine several flat counters in a single package and thus increase the number of radiator layers.

The flat counter which we assembled consisted of two thin Plexiglas disks covered with 0.01 mm thick copper foil. Tin dioxide enriched to 75% of the isotope Sn^{119} was deposited on the foil. The thickness of the front layer was 0.5 mg/cm^2 and the back layer, 1 mg/cm^2. In total, these thicknesses give the value $C_r = 2$.* Three well-drawn filaments of 0.1-mm tungsten wire were placed between the two facing surfaces. One filament passed exactly along the diameter of the counter, and the two others were located at a distance of 10 mm on either side of it. The diameter of the working surface of the counter was 40 mm, and the distance between the disks was 5 mm. Leads from both disks and all the filaments were brought to the outside. The counter, which was filled with a mixture of 80% Ar and 20% ethylene, was tested separately with voltage supplied to the central filament, the two side filaments, and to all three filaments. The counting characteristics had the same form for all three methods of supply. With two wires the count was 30% greater, and with three wires 50% greater than with one wire. During operation with one wire the length of the plateau was somewhat greater than in the other cases, rising less than 10% in 150 volts.

The counter was tested on apparatus with a linearly variable velocity, and the absorption spectrum was recorded on a 100-channel analyzer /7/.

---

* In determining C_r we used the value $f' = 0.46 \pm 0.3$, which we measured for SnO_2.

The source used was in the form of the compound SnO_2. The absorption spectra obtained with the resonance counter were compared with spectra measured with a scintillation spectrometer using the same geometry.

An absorption curve measured by the resonance counter with the setup in Fig. 1a is given in Fig. 4. The counting maximum here corresponds to a complete overlap of the emission and absorption lines when v = 0.

The scintillation counter was used to determine the background line (broken line in Fig. 4), which in this case was not straight because different values of the velocity were realized at different source-to-counter distances.

Since resonance is completely destroyed when the source is moving at high speed, and the count in this case is caused by the recording only of "nonresonance" γ-rays, the amount by which the resonance count exceeds background can be precisely determined from these measurements as follows:

$$\frac{N_r(0) - N_b(0)}{N_b(0)} = 250\%.$$

Then the source was fixed relative to the counter and further work was done by the transmission method with the setup in Fig. 1b. Absorption spectra were measured with a 22-mg/cm^2 SnO_2 absorber (Fig. 5). The relative magnitudes of the effect were: $\eta_r = 60\%$ for the resonance counter and $\eta_0 = 25\%$ for the scintillation counter. From a comparison of the absorption line intensities we obtained a value of 15% for the efficiency of the resonance counter for recoilless γ-quanta.

The same sort of measurements were made without the palladium filter which is always used with a scintillation counter in order to attenuate the 25 keV x-rays of Sn (Fig. 6). As was to be expected, the absorption line measured with the scintillation counter was considerably weaker, $\eta_0 < 12\%$, because of the increase in background from the x-rays, whereas with the resonance counter the magnitude of the effect was practically unchanged. An estimation of the efficiency of the counter for x-rays based on these results gives a value less than 0.1%. These measurements showed that in using a resonance counter it is possible to do away with the palladium filter and thus increase the counting intensity by 2 to 3 times.

All absorption line widths measured with the resonance counter were somewhat less than the widths obtained by the usual method of recording.

The difference in widths can be attributed to the selective sensitivity of the resonance detector. In order to test what effect emission line broadening in a thick source has on the absorption line width, which should be slight for a resonance detector, it would be necessary to make measurements with sources of different thickness. However, such sources were not at our disposal, and the lines were artificially broadened with the aid of an auxiliary fixed absorber of SnO_2 with a thickness of 12 mg/cm^2. The absorption spectrum was measured with a moving, 11-mg/cm^2 absorber. With the scintillation counter we obtained a line of width $\Gamma' \simeq 3$ mm/sec with $\eta_0 = 8\%$, whereas with the resonance detector a line having $\Gamma = 1.5$ mm/sec and $\eta_r = 30\%$ was obtained (Fig. 7).

Thus a considerable decrease in the effect of the source line width on broadening of absorption spectra has been experimentally demonstrated. It should be noted that the absorption lines given in Fig. 7 were measured for the same length of time; however, in spite of a considerably greater count, the line of spectrum 1 is not clear, whereas spectrum 2 forms a well-marked dip corresponding to the absorption maximum.

The following conclusions can be drawn from the experiments performed:

1. The efficiency of a resonance counter with an SnO_2 radiator for 23.8-keV recoilless γ-rays is two orders of magnitude greater than its efficiency for x-rays of the same energy.

2. As a result of the advantageous recording of recoilless γ-rays a resonance detector makes it possible to combine a high counting rate with an effect of considerable magnitude.

3. The use of a resonance counter in transmission experiments permits a decrease in line width. This decrease is particularly significant in working with thick sources.

4. It should be noted that the use of a resonance counter makes it possible to simplify the recording apparatus considerably and to increase the stability of its operation.

Institute of Nuclear Physics, Moscow State University

REFERENCES

/1/ K. P. Mitrofanov and V. S. Shpinel'. ZhÉTF 40, 983 (1961).

/2/ E. Kankeleit. Z. Physik 164, 442 (1961).

/3/ V. A. Bryukhanov, N. N. Delyagin, V. Zvenglinskii, V. S. Shpinel'. ZhÉTF 40, 713 (1961).

/4/ S. Margulies and F. Ehrman. Nucl. Instr. 12, 131 (1961).

/5/ D. Shirley, M. Kaplan and P. Axel. Phys. Rev. 123, 816 (1961).

/6/ G. A. Bykov, Pham Zuy Hien, V. S. Shpinel'. In press.

/7/ N. N. Delyagin, V. S. Shpinel', V. A. Bryukhanov. ZhÉTF 41, 1347 (1961).

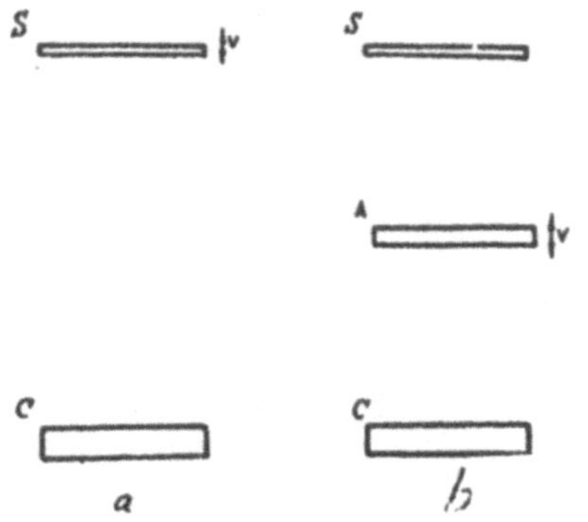

Fig. 1. Setup for making measurements on the resonance scattering of γ-rays. S) Source; C) counter; A) absorber. The symbol ↕ indicates an object in motion.

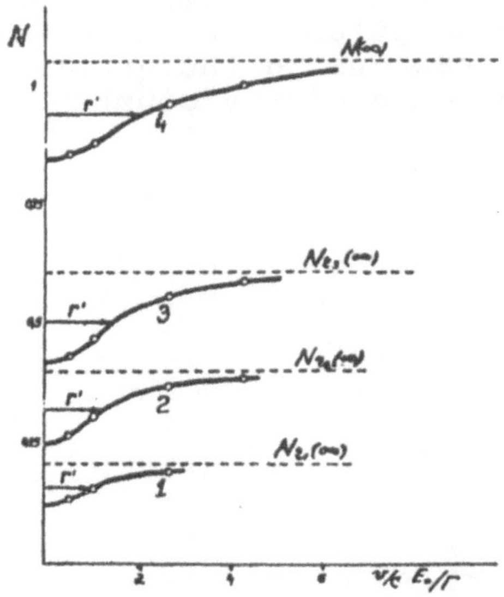

Fig. 2. Absorption spectrum curves calculated for the case $b = 3$, $C_a = 1$, with the assumption that all γ-quanta are emitted without recoil. The lines 1, 2, and 3 correspond to measurements by resonance counters with thicknesses $C_r = 1$, 3, and 10, respectively. Line 4 was calculated for recording with an ordinary detector. $N(\infty)$ is the counting level without consideration of resonance absorption in the absorber.

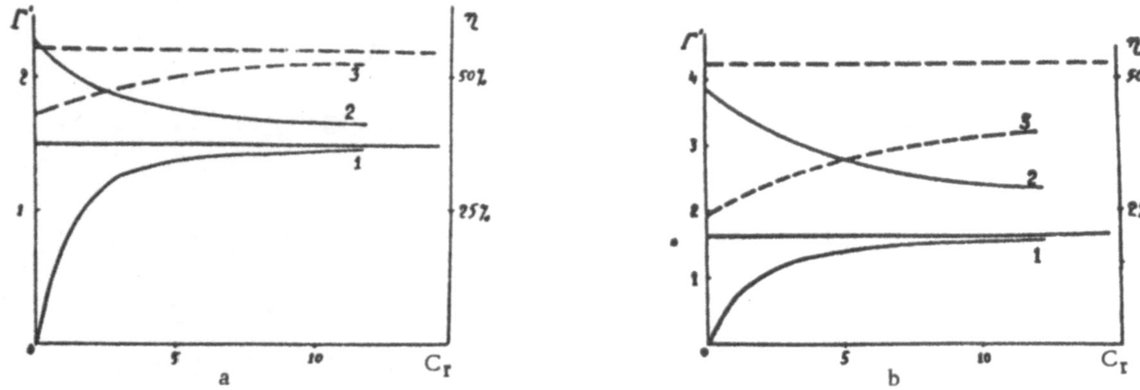

Fig. 3. Magnitude of effect and absorption line width as functions of radiator thickness in a resonance counter. Curve 1) η_0 is the absolute magnitude of the effect (the fraction of γ-quanta absorbed in the absorber with $\varepsilon = 0$, relative to the total flux of γ-quanta). Curve 2) η_r is the relative magnitude of the effect (the ratio of the number of γ-quanta absorbed in the absorber to the counting level of the given detector when $v = \infty$). Curve 3) Γ' is the width of the absorption line in multiples of Γ. The horizontal straight lines correspond to the quantities η (solid line) and Γ' (broken line) for an ordinary counter.

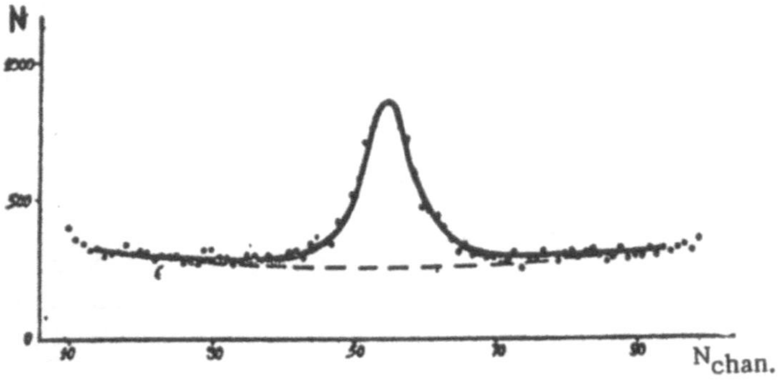

Fig. 4. Resonance absorption spectrum measured with a counter in the setup of Fig. 1a. The broken line indicates the background.

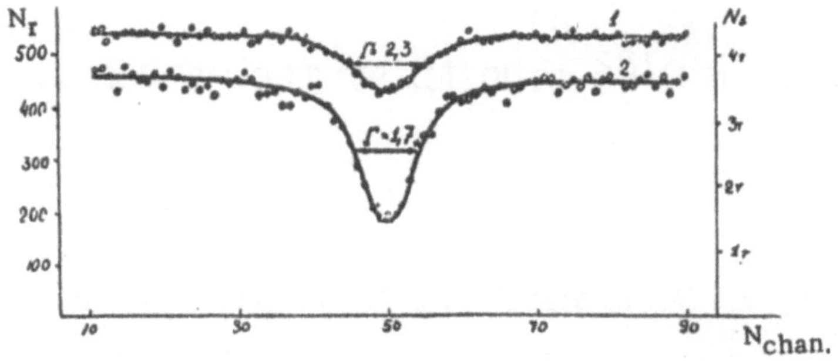

Fig. 5. Absorption spectrum measured with 22-mg/cm^2 SnO_2 absorber. Curve 1) Scintillation spectrometer; curve 2) resonance counter.

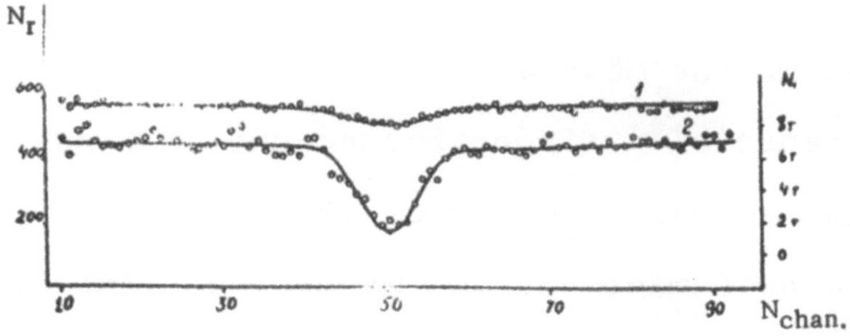

Fig. 6. Resonance absorption spectrum measured without palladium filter in front of source, with 22-mg/cm^2 SnO_2 absorber. Curve 1) Scintillation spectrometer; curve 2) resonance counter.

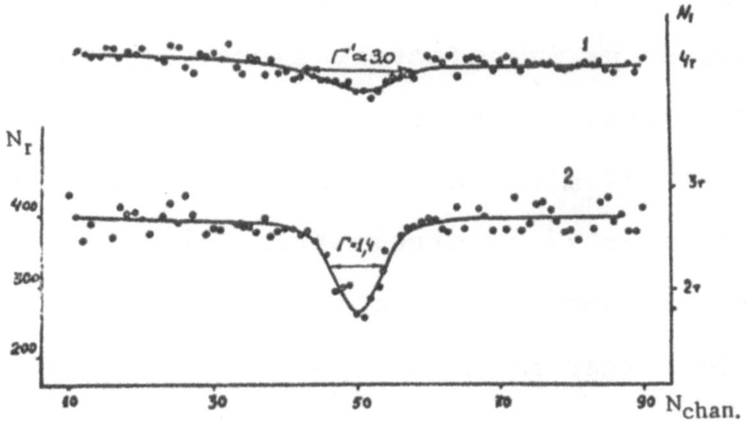

Fig. 7. Resonance absorption spectrum measured with artificial broadening of emission line from source. Curve 1) Scintillation spectrometer; curve 2) resonance counter.

Translated by S. Elliott

THE MÖSSBAUER EFFECT IN W^{182} STUDIED BY THE METHOD OF γ-γ COINCIDENCE

Tsv. Bonchev, L. Mitrani, Sl. Ormandzhiev, B. Skorchev, Iv. Uzunov

Observation of the Mössbauer effect /1/ in W^{182} is made difficult by the presence of a strong background of lines with energies greater than the resonance energy of 100 keV. Attempts to record this effect have been made by Lee, Meyer-Schutzmeister, Schiffer, and Vincent /2/ and by de Nercy, Langerin, and Spighel /3/, who studied changes in γ-ray absorption resulting from changes in temperature of the source or the absorber. Both of these attempts cannot be considered as satisfactory since, for the above-mentioned reason, the effect under observation is very slight. Recently this difficulty has been eliminated by Sumbaev, Smirnov, and Zykov /4/ by using a curved-crystal spectrometer. Since this last method involves complicated and hard-to-obtain apparatus, there is good reason to look for other methods. This, then, is the purpose of the present work.

BASIC PRINCIPLES OF THE PROPOSED METHOD

Instead of resonance absorption of the radiation passing through the absorber, we shall use resonance scattering /5/, in which case the effect will be observed as differences in scattering caused by variation of the temperature of the source. Since

Compton photons, primarily from the lines at 152, 222, 229, and 264 keV, fall into the 100-keV region being investigated, we shall exclude the possibility of counting them by using coincidence techniques. This is possible because the listed γ-lines are associated in cascade with the transition for which resonance absorption is observed, while the lines themselves are not related to one another in cascade (Fig. 1) /6/. From the decay scheme it can be seen that only γ-quanta with an energy of 229 keV could give a noticeable number of coincidences with γ-quanta of the 222- and 264-keV transitions, but these coincidences will be a small percentage of the total number of coincidences.

METHOD OF MEASUREMENT

The setup of the apparatus is shown in Fig. 2. The source is mounted on a thin platinum strip serving as a heater and attached to a 2 mm thick Pertinax plate, which, in turn, is mounted on a 2 mm thick aluminum plate connected with the liquid nitrogen vessel. The scatterer, a tungsten wire wound on a thin aluminum cylinder, is located to one side of the source. The scatterer is surrounded on all sides by liquid nitrogen. An absorber, which shields the scintillation counter SC-I from the direct γ-radiation, is situated in front of the source. The source temperature is measured by a Cu-constantan thermocouple.

Using the fast-slow coincidence scheme shown in Fig. 3, coincidences are obtained between γ-quanta which are recorded by the scintillation counter SC-II located directly behind the source and quanta recorded by the scintillation counter SC-I, which records scattered radiation. In this way γ-quanta in the 100-keV region which are incident on the crystal of SC-I are recorded only if a γ-quantum of the transitions 152, 222, 229, or 264 keV is incident on the crystal of SC-II at the same instant. Thus we have removed the background of Compton radiation with an energy which coincides with the energy of the resonance transition but which is caused by quanta of higher energy.

In order to reduce the loading of SC-II we placed a filter of lead and tin plates between it and the source, which sharply reduced the intensity of the strongest spectral line at 68 keV and the 100-keV line in comparison with the lines at 152, 222, 229, and 264 keV (Fig. 4). In order that the 100-keV line be better isolated in the SC-I channel, on the other side of the source we used a cadmium filter, which absorbs 68-keV quanta several times more strongly than 100-keV quanta. Under these circumstances the 100-keV line, which in a direct spectrum is more weakly separated than the lines around it, is rather clearly isolated in the coincidence spectrum.

A direct spectrum and a coincidence spectrum, which were made under identical conditions with no filter, are shown in Fig. 5.

The expediency of this method can be seen best from Fig. 6, where the spectra of radiation scattered by tungsten are shown with and without coincidences. It is evident that by the use of coincidences the strong Compton background in the vicinity of 100 keV is decreased by many times, which permits the appearance of the resonance scattering effect.

Since the experiment was performed with comparatively small solid angles, the number of coincidences was small -- about 600 in 1000 sec. This fact imposes considerable requirements on the apparatus. In order to stabilize the operation of the amplifiers they were wound with heaters and their temperatures held constant with an accuracy of 2-3 C°. The background of accidental coincidences was reduced practically to zero by choosing the activity of the source to correspond with the resolving power of the fast-coincidence circuitry (about 0.5×10^{-8} sec).

MEASUREMENTS

Two series of measurements were carried out. The relative change in scattering, i.e., the quantity

$$[N_{t=-196°C} - N_{t=20°C}]/N_{t=-196°C}$$

as the source varied from liquid nitrogen temperature to room temperature, was determined by a student to have the same value of 9.5 ± 3.5 % (with a reliability of 75%).

The fact that in this case we are not dealing with changes in intensity of radiation due to mechanical deformations during heating of the source was established by a parallel recording of the total number of pulses from SC-I, which remained constant within the limits of statistical error.

CONCLUSION

The results obtained with the aid of the described apparatus show that there is the possibility of considerably increasing the directly recordable Mössbauer effect in W^{182} by means of γ-γ coincidences. In /4/ the effect was about 2.5%. This method could prove to be useful also in other cases.

We are continuing our investigations, with the main problems being increasing the solid angle within which the scintillation counters record radiation and also stabilizing the electronic apparatus.

Sofia State University

REFERENCES

/1/ R. L. Mössbauer, Z. Physik 151, 124 (1958).

/2/ L. L. Lee, L. Meyer-Schutzmeister, J. P. Schiffer, and D. Vincent, Phys. Rev. Letters 3, 5, 223 (1959).

/3/ A. Bussiere de Nercy, M. Langerin, and M. Spighel, J. phys. radium 21, 288 (1960).

/4/ O. I. Sumbaev, A. I. Smirnov, V. S. Zykov. ZhETF, 42, 115, 1962.

/5/ R. Barloutaud, J.-L. Picou, and C. Tzara, Compt. rend. 250, 2075 (1960).

/6/ O. Fröman, M. Ryde, Arkiv for Fysik 12, 399 (1957).

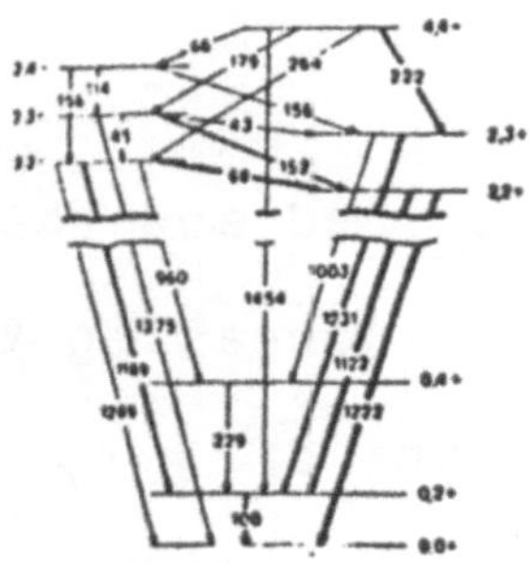

Fig. 1. Decay scheme of Ta^{182}.

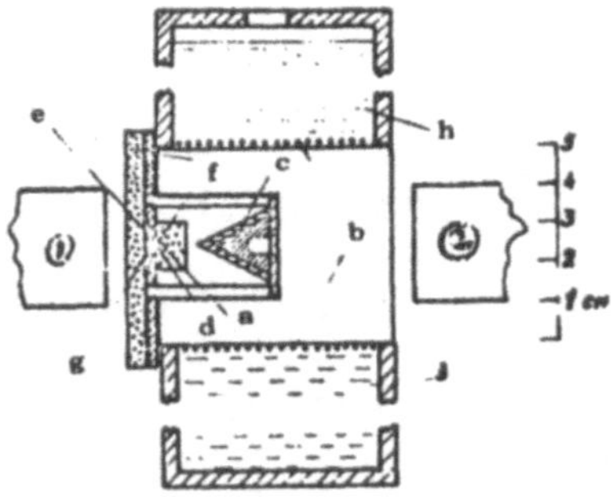

KEY:
1) SC–II
2) SC–I

Fig. 2. Setup of apparatus. a) Source; b) scatterer; c) absorber; d) heater; e) thermocouple; f) thermal insulator (Penoplast); g) Pertinax plate; h) liquid nitrogen.

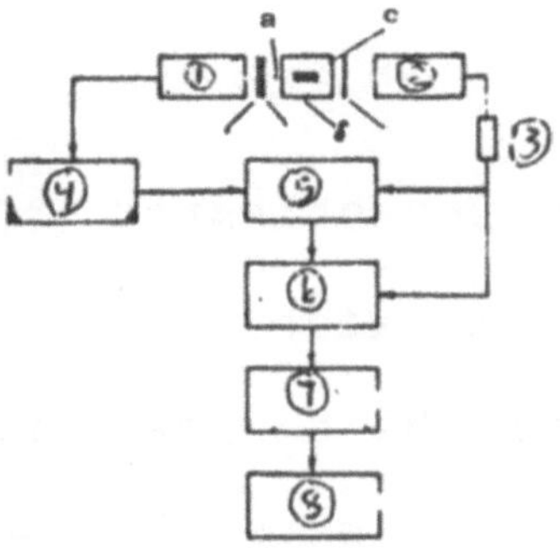

Fig. 3. Block diagram of electronic equipment.
a) Source; b) scatterer; c) inert absorber.

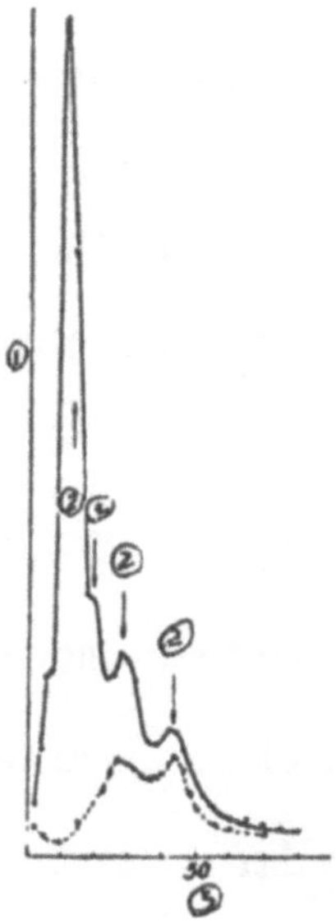

Fig. 4. Direct spectrum of Ta recorded on scintillation counter SC-II.

—— without filter
--- with filter (Pb and Sn)

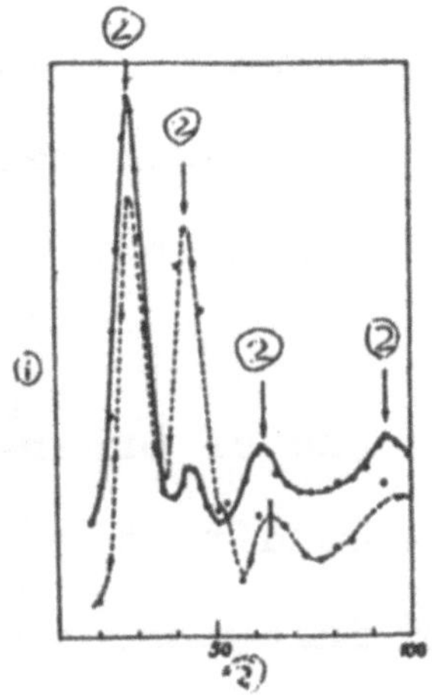

Fig. 5. Spectrum of Ta^{182} in low-energy region.

—— direct spectrum
--- coincidence spectrum

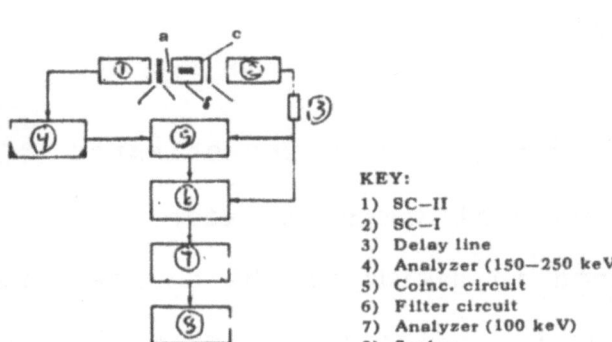

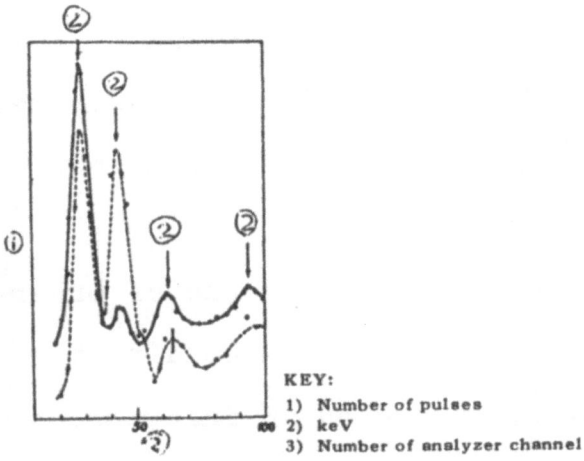

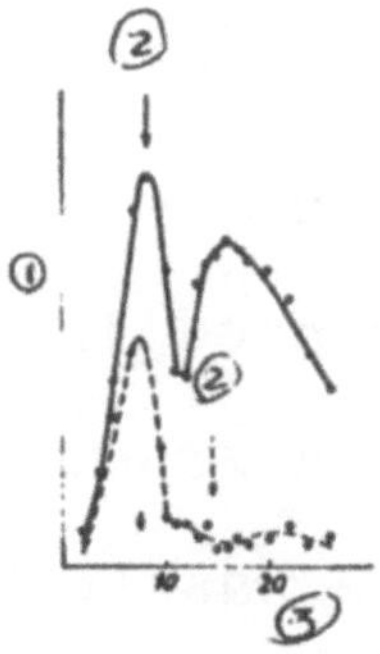

Fig. 6. Spectrum of radiation scattered by tungsten.

—— direct spectrum
--- coincidence spectrum

Translated by S. Elliott

RECOILLESS RESONANCE γ ABSORPTION IN Sm^{149}

V. P. Alfimenkov, Yu. M. Ostanevich, T. Ruskov

A. V. Strelkov

Resonance absorption of 22.5 keV γ's has been investigated, with the source and absorber made of natural samarium oxide.

The radioactive mother nuclei used in making up the source were introduced as $Eu_2^{149}O_3$ in a thin layer (~ 5 mg/cm^2) of natural samarium oxide. The absorber was 50 mg/cm^2 thick. The powdered materials in source and absorber were fastened to the backing with polyvinyl alcohol to provide mechanical strength.

The resonance absorption spectrum was obtained in the usual way from the Doppler shift, using an electromagnetic vibrator.

The electromagnetic vibrator (Fig. 1) consisted of two dynamic loud speaker units mounted together. The two windings, rigidly held together by an aluminum rod, were centered in the annular magnet gaps by two diaphragms.

The absorber, placed in a thin walled container made of organic glass, was mounted on a frame fastened to the rod connected to the windings. The source was held inside the frame on an independent suspension, and remained fixed during the measurements.

After passing through the absorber, the γ-radiation was recorded on an argon proportional counter. The γ energy range that we were interested in was isolated from the amplitude spectrum with an AADO-1 single channel analyzer. The vibrator windings and the absorber moved according to a cosine law at the natural frequency of the vibrator. The vibrator that we used made it possible to get rates of absorber motion from fractions of a millimeter to tens of centimeters per second.

The scheme (Fig. 2) was used to excite vibrator oscillations at the resonant frequency with a sufficiently stable amplitude controllable over a wide range, and to get the relation between the counting rate and the rate of motion of the absorber. There was a triggering pulse generator with a frequency 100 times greater than the natural frequency of the vibrator, where the generator frequency could be varied within definite limits by changing the external voltage. The pulses from the generator went through the input of a scaling circuit with a scaling factor of 100, which gave a rectangular voltage wave at the output. After power amplification, this voltage was fed to one of the vibrator windings. The voltage induced by the motion in the other winding, after nonlinear amplification to form a rectangular wave, was fed to the phase meter along with the voltage supplying the first winding. The D.C. voltage from the phase meter output controlled the generator frequency, thus giving inverse frequency feedback. This system gave satisfactory long time stability of the vibrator oscillations at the natural frequency.

The outfit had provisions for stroboscopic monitoring (in conjunction with a microscope) of the law of motion followed by the vibrator. The pulses from the generator (without scaling) were fed to the address device of a standard AI-100 amplitude analyzer, and switched in all 100 analyzer channels during one period of the vibrator oscillations.

After passing through the blocking and forming systems, the

pulses from the single channel analyzer output were fed to the arithmetic device of the AI-100 analyzer, and were recorded in the channels corresponding to their time of arrival.

Switching the channels at equal time intervals with a harmonic motion law gave the resonance absorption spectrum to a nonlinear energy scale.

Means were provided for controlling the time shift of the channels with respect to the phase of the vibrator motion. However, this control was only used from considerations of convenience, since there was no trouble about deciphering the picture for the channels in any arbitrary position with respect to the phase of the motion. Actually, during one period of the motion, all the velocities are gone through twice with a time displacement of exactly a half period (Fig. 3). This means that the spectrum in the analyzer channels consists of two mirror-image symmetric parts, with the zero velocities lying 25 channels on either side of the point where the symmetric parts of the spectrum come together.

The measurements, made on the equipment, showed that the resonance absorption spectrum of Sm^{149} in samarium oxide, up as far as velocities of 2 cm/sec ($\Delta = 1.5 \cdot 10^{-6}$ eV) contained only one practically unshifted component.

The results of one series of measurements are shown in Fig. 3.

The curve shown in the figure was obtained from a least squares analysis of the experimental data on an electronic computer.

The analysis gave the following parameters for the resonance absorption spectrum:

$$\Gamma = (1.53 \pm 0.1)\cdot 10^{-7} \text{ eV}$$

$$\Delta E = (0.19 \pm 0.12)\cdot 10^{-7} \text{ eV}$$

$$\varepsilon = (10 \pm 1)\cdot 10^{-2}, \text{ where}$$

Γ is the width of the γ- transition investigated, ΔE is the shift between the emission and absorption spectra, and ε is the effect at zero velocity including background correction.

Measurements of ε for several natural filters with thicknesses less than 50 mg/cm^2 have shown that the filters are thin for thicknesses up to 50 mg/cm^2. This prevented us from finding the values of P and P', since, with the unfavorable shape of the spectrum, and our comparatively weak source, it was difficult to make measurements with unenriched filters of large thickness. The measurements on the Mössbauer effect in Sm^{149} will be continued.

In conclusion, the authors thank F. L. Shapiro for his interest in the work, as well as N. A. Lebedev and A. F. Novgorodov for making up the source.

The United Institute of Nuclear Studies

Translated by: Charles V. Larrick

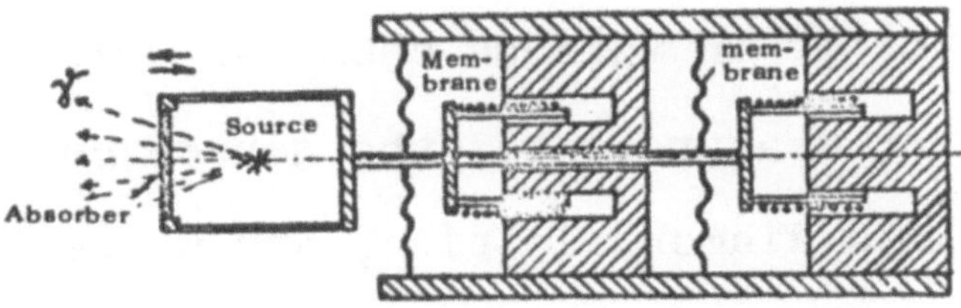

Fig. 1

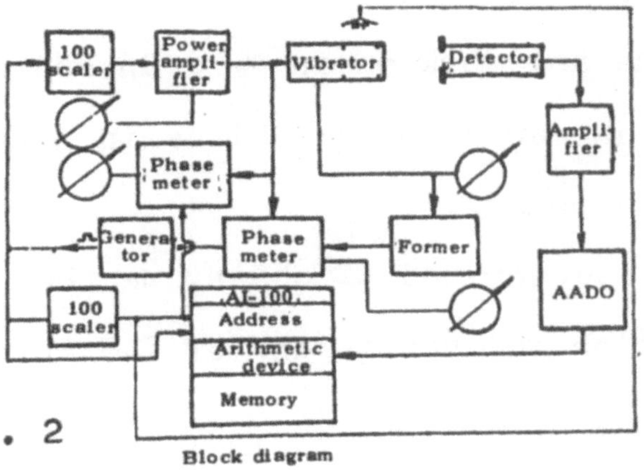

Fig. 2

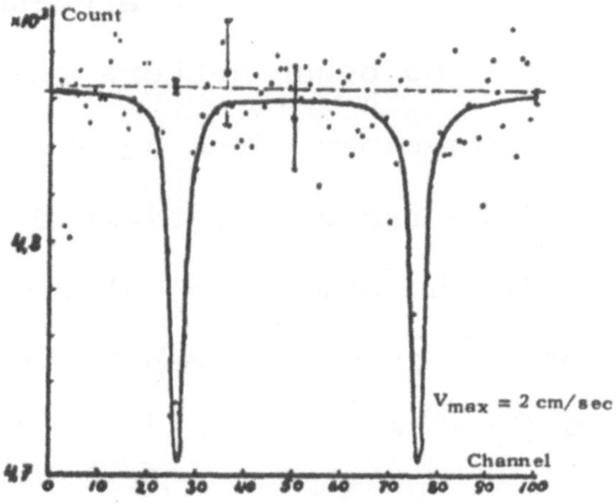

Fig. 3

THE MÖSSBAUER EFFECT IN Sn^{119} NUCLEI, ATTEMPT TO OBSERVE THE EFFECT IN Pr^{141}

V. A. Bukarev

INTRODUCTION

A large amount of work has been devoted to the study of the Mössbauer effect in various tin compounds. However, the experimental results of a number of authors on various compounds are widely different. For example, in /1--5/, clearly resolved quadrupole splitting was found in white tin, while other papers /6,7,8/ give a value several times less for the upper limit of this quantity. The conclusions of /5/ seem rather strange in saying that there is absolutely no electric field gradient (quadrupole splitting) at the tin nuclei in the tetragonal SnO_2 lattice.

In order to clear up these and other contradictions, a study has been made in the present paper of resonance absorption of 23.8 keV γ-radiation by Sn^{119m} in various chemical compounds of tin.

In this series of measurements, the source used was gray tin, which crystallizes in a diamond lattice, and where splitting of the emission line is to be expected, as confirmed in /10/. As far as we know, no other authors have done any work on an α - Sn source.

This paper gives data on the isomeric energy shifts in a number of so far uninvestigated tin compounds.

At the same time, attempts are described at observing the Mössbauer effect from both absorption and scattering of 145 keV γ's from Ce^{141} by Pr^{141} nuclei.

EXPERIMENTAL SETUP

In the resonance absorption work, the source and absorber were placed in pentoplast cryostats, and could be cooled to liquid nitrogen temperature. The experiments described below, unless otherwise specified, were made at this temperature. The absorber cryostat, placed between the radiation detector (FÉU-29 scintillation counter with NaI crystal) and the source cryostat, was moved back and forth with an excentric.

The radiation was only recorded during the parts of the motion which corresponded with a constant translational velocity. During one measurement cycle, nine relative velocities were taken in succession, including zero velocity, the same for all cycles. For this purpose, the field voltage was automatically changed at definite intervals on the D.C. motor, which drove a cam mechanism through a rigid set of reducing gears.

With the velocities set ahead of time with a special device, and using a variable reducing gear, the resonance absorption spectrum could be gone over in detail at any part, over a range of ± 8 mm/sec. The velocity instability was not greater than ± 0.15 mm/sec in the range 1--8 mm/sec, and was 0.05 mm/sec for velocities below 1 mm/sec.

The spectrum in the resonance radiation range was isolated with the single channel amplitude analyzer (AADO-1). The electrical pulses for a definite rate of motion were recorded in a definite counting channel of the equipment. In order to avoid errors coming from instability of the mechanism fixing the recording time at different velocities and controlling the switching, separate time channels were put into the equipment. These channels gave strict statistical control of when the time was taken.

The results of the measurements were reduced to indications at zero relative velocity, which reduced the effect of slow changes in the apparatus.

Experiments with tin compounds.

The outfit described above was used to take the resonance absorption spectra* of 23.8 keV γ's by Sn^{119m} in 13 different chemical compounds of tin. Two types of sources were used in the measurements. One was the α- modification of tin, while the other was the compound SnO_2. The sources were made of tin foil, enriched to 94% Sn^{118}, irradiated in a reactor. The first type of source was prepared by keeping one of the foils for a long time ($\sim$15 days) at dry ice temperature with a crystallite of gray tin present as a catalyst. The second type was prepared chemically with subsequent annealing at 1200--1400° C for 30 hours to restore the crystal structure.

Figure 1 illustrates the absorption spectra in β-Sn with a thickness of 9.9 mg/cm^3 (a) and in $Na_2SnO_3 \cdot 3H_2O$, 10 mg/cm^2, by Sn (b), from a gray tin source. The abscissas are the velocity of the absorber relative to the source in mm/sec (the "+"sign means approaching), while the ordinates are the resonance absorption effect in percent of the counting rate at zero velocity. It may be seen from the curves that the half widths are 0.85 and 0.9 mm/sec respectively, which is sufficiently close to twice the natural line

*The radiation intensity as a function of the relative velocity between the source and the absorber.

width, 0.62 mm/sec, found from the half life of the first excited level (1.85 $\pm$ 0.1) 10^{-8} sec /9/.

Reducing the absorber thicknesses to 5 mg/cm^2 did not, within the error of measurement, produce any appreciable change in the resonance line widths.

A very substantial broadening of the absorption curves in the same β- Sn and $Na_2SnO_3 \cdot 3H_2O$ samples of up to 1.4--1.5 mm/sec occurred when using an SnO_2 source. Figure 2 shows the Mössbauer spectrum in $Na_2SnO_3 \cdot 3H_2O$, 10 mg/cm^2 thick. Our attention is called to the absence of isomer shift between these compounds.

Quadrupole interaction between the Sn^{119} nucleus and the crystal lattice showed up clearly in the bivalent tin compounds $SnCl_2 \cdot 2H_2O$ and $SnSO_4$, the spectra of which are shown in Fig. 3*. Using $SnSO_4$ absorber with an α-source gave a narrower line with a clearly defined flat top. However, having no resolved peaks makes it difficult to get an accurate value for the splitting Δ. For $SnCl_4 \cdot 2H_2O$, Δ is 1.25 $\pm$ 0.15 mm/sec.

A very considerable Mössbauer effect was observed in compounds with low melting points, and light atoms in the crystal lattice. For example, in $SnCl_4 \cdot 5H_2O$ (40 mg/cm^2), the effect is about 28% at the maximum. The theoretical questions of Mössbauer line intensity in the presence of an impurity atom are discussed in /11--14/.

The data on the isomeric energy shifts in various tin compounds relative to the γ- transition in the SnO_2 source are given

*The resonance effect in $SnCl_2 \cdot 2H_2O$ refers to the counting rate at negative velocities.

in the table.

DISCUSSION OF RESULTS OBTAINED

The experimental data was used to calculate $\underline{f}'$, the probability of resonance γ- capture by Sn^{119} nuclei in various crystal structures. Two methods were used for the purpose.

If there is no quadrupole splitting, or it is not large in comparison with the line width, $\underline{f}'$ may be found uniquely from the ratio of the area under the curve to the resonance absorption effect at the maximum, ε. This eliminates the resonance characteristic of the source, αf.

The limits of variation of the two unknown parameters $\underline{f}'$ and αf are also found from the intersection of the curves

$$\varepsilon = \alpha f\left[1 - I_0\left(\frac{a}{2}\right)e^{-\frac{a}{2}}\right]$$

for several absorbers of different thickness. Here $I_0(a/2)$ is the zero order Bessel function of imaginary argument, $a = \sigma_0 f' n$, σ_0 is the cross-section at resonance, and $\underline{n}$ is the number of nuclei of the resonant isotope per cm^2 of absorber.

At liquid nitrogen temperature, the averaged values of $\underline{f}$ for β-Sn and $Na_2SnO_3 \cdot 3H_2O$, found by different methods, turned out to be equal respectively to 0.35 ± 0.07 and 0.56 ± 0.06. The resonance parameter αf was 0.28 ± 0.03 for an α-Sn source.

Calculating the broadening of the α- source emission spectrum from the self absorption showed that the half width cannot increase by more than 0.05 mm/sec. There was no appreciable effect of vibrations on the deformation of the curves in the working range of velocities. Obviously, these difficulties lie in errors of measurement.

Accordingly, the broadening of the absorption curves observed for β-Sn and $Na_2SnO_3 \cdot 3H_2O$ can, basically, be accounted for by a small amount of quadrupole splitting in the compounds, since the half waves remain practically unchanged with change in absorber thickness.

Possible limits for the value of Δ in β-Sn are given in /17/ as $0.19 \leqslant \Delta \leqslant 0.5$ mm/sec.

From our data, the upper limit must be reduced to 0.35 mm/sec. This value is not in contradiction with /6,7,8/, but is clearly at variance with the conclusions of /1--5/, where a well resolved splitting was observed, equal to 1.4 mm/sec.

Considering what has been said above, it may be assumed that the effect of finite absorber thickness cannot be wholly responsible for the broadening of the resonance spectrum even in $Na_2SnO_3 \cdot 3H_2O$, even if we take into consideration some difference between the values of <u>f</u>' for β- Sn and $Na_2SnO_3 \cdot 3H_2O$. There is apparently quadrupole splitting in sodium stannate, with the upper limit at 0.4 mm/sec.

The absence of any isomeric energy shift between tin dioxide and sodium stannate indicates that the Sn^{119} nucleus is in identical environments in these compounds. On the other hand, the very substantial spectrum broadening of up to 1.4 mm/sec forces us to assume that there is splitting even in SnO_2. An analysis of the data obtained when using SnO_2 absorbers (at room temperature and liquid nitrogen temperature) with α-Sn and SnO_2 sources, has shown that the upper limit of Δ in SnO_2 is much larger than the splitting in $Na_2SnO_3 \cdot 3H_2O$.

This conclusion is obviously in contradiction with the conclusions of /4,5/ that quadrupole splitting is completely absent in the tin nuclei in the SnO_2 lattice. It should also be noted in this connection that no satisfactory agreement has been obtained either from trying to plot the experimental half widths for tin dioxide from an SnO_2 source (absorber and source at room temperature) on the graph of $\Gamma_{exp} = f(n)$ given in /4/, where the small amount of extrapolation to zero absorber thickness, made by the authors, gives a value that is practically the same as twice the natural width of the excited level. A particularly large difference is observed for thin absorbers. For example, for a thickness of 5 mg/cm^2, $\Gamma_{exp} = 10 \cdot 10^{-8}$ eV, while the above authors give a value 30% less.

The isomeric shifts in the tin compounds studied divide, as it were, into two groups, corresponding to bi- and quadrivalent structures, with the exception of $SnBr_2$. Assuming a purely ionic structure for the compounds, we can make an approximate calculation of the lower limit of change in the effective radius of the nuclear charge in the excited state, $\Delta R/R$. Considering the external electron configuration of the tin atom, $5s^2 5p^2$, and assuming that the 5s electron density in the nucleus is equal to zero in quadrivalent compounds, we can, from the difference in energy of the transitions in SnO_2 and $SnSO_4$, find $\Delta R/R \geqslant +0.8 \cdot 10^{-4}$*. This quantity, calcu-

---

*The value of $\Delta R/R$ was reported on by the author at the Conference on Nuclear Spectrometry held in Leningrad in February 1962.

lated with 30% accuracy in the same way in /10/, turned out to be equal to $+1.1 \cdot 10^{-4}$.

Figure 4 shows the results of measuring the energy shifts of the excited level in various tin compounds relative to the transition in SnO_2, compared with the results of the Scientific Institute for Nuclear Physics Research group at Moscow State University /5/ *. It may be seen that the isomeric shifts are in good agreement when the absorbers are the same. The same figure shows the results obtained by Boyle et.al. /10/. These data, since the measurements were made with a β^- source, are plotted relative to the intermediate value of the shifts for white tin, which we found in /5/. Our attention is called to the large difference, beyond the limit of measurement error, in the isomeric shifts for $SnCl_2 \cdot 2H_2O$. Further, no quadrupole splitting was found for this compound in /10/, while it showed up clearly in our measurements (see Fig. 3a). In subsequent checks with $SnCl_2 \cdot 2H_2O$ absorbers made up in different ways, a well resolved splitting and substantially less isomeric energy shift were always observed.

It should be noted that there is too much difference between the isomeric shifts for SnF_2 and $SnCl_2$ found in /5/ and /10/, and that it is beyond the limits of error in measurement.

EXPERIMENTS WITH Pr^{141}

In addition to the tin experiments, some attempts were made to observe the Mössbauer effect in Pr^{141} nuclei. A study was made

---

*The measurements were made with an SnO_2 source at room temperature.

of the 145 keV γ-transition $7/2+ \longrightarrow 5/2+$ with an excited level half life of $(2.0 \pm 0.3) \cdot 10^{-9}$ sec /9/, which corresponds to twice the natural width of the spectrum, 0.94 mm/sec. The emitter was the radioactive isotope Ce^{141} in the compound CeO_2. Cerium dioxide crystallizes in a cubic lattice, and obviously no quadrupole splitting of the emission line is to be expected. However, as we know, all the rare earth elements and their compounds are strongly paramagnetic substances. It is accordingly possible to have Zeeman splitting in both the source and the absorber.

The absorber was the complex compound $Pr_2O_3 \cdot PrO_2$ ($\sim$900 mg/cm^2 in Pr).

No resonance effects could be observed within $\pm 0.2\%$ in absorption measurements made in the velocity range ± 8 mm/sec.

An attempt was then made to investigate this absorber for a 90° scattering effect. In this experiment, the scatterer was at room temperature.

The multipolarity ratio of the emitter, E2/M1, for the 145 keV γ-transition is 0.007 ± 0.003, and the conversion coefficient is 0.3 /9/. The anisotropy in the angular distribution of the resonantly scattered quanta, calculated for the pure M1 of the transition, turned out to be not more than 20%.

Since collimation could not be used in the last case to eliminate the difference in counting geometry with the absorber fixed for even some velocity, the results of the measurements were referred to the data on scattering by lanthanum oxide of approximately the same thickness. During the experiment, the scatterers were alternately exchanged.

However, even here no effect was observed to $\pm$ 1.5%.

In conclusion the authors recognize their pleasant duty of expressing sincere gratitude to F. L. Shapiro for his daily interest and great help in the work.

N. P. Lebedev Physics Institute, Academy of Sciences, USSR

Translated by: Charles V. Larrick

TABLE

Contents	Shift, MM/sec
SnO_2	0
$Na_2SnO_3 \cdot 3H_2O$	0
$SnBr_2$	0,20±0,05
$SnCl_4 \cdot 5H_2O$	0,25±0,05
$(NH_4)_2SnCl_6$	0,50±0,I0
$SnCl_4$	0,85±0,I0
$SnBr_4$	I,I5±0,25
α-Sn	2,0 ±0,I0
β-Sn	2,55±0,I5
$SnCl_2 \cdot 2H_2O$	3,55±0,I5
$SnSO_4$	3,95±0,I5
$SnCl_4 \cdot 2\{C_2H_4(NH_4)_2\}$	0,25±0,05
$[SnCl_6]^{-2}\ 2\alpha$-An	0,40±0,I0

The complex tin compounds were studied in cooperation with V. I. Gol'danskii's laboratory.

REFERENCES

/1/ N. N. Delyagin, V. S. Shpinel', V. A. Bryukhanov, B. Zvenglinskii, ZhÉTF, 39, 220 (1960)

/2/ V. A. Bryukhanov, N. N. Delyagin, B. Zvenglinskii, V. S. Shpinel'. ZhÉTF 40, 713 (1961)

/3/ V. S. Shpinel', V. A. Bryukhanov, N. N. Delyagin, ZhÉTF 40, 1525 (1961)

/4/ N. N. Delyagin, V. S. Shpinel', V. A. Bryukhanov, ZhÉTF 41, 1347 (1961)

/5/ V. S. Shpinel', V. A. Bryukhanov, N. N. Delyagin, ZhÉTF 41 1767 (1961)

/6/ A.T.F. Boyle, D. St. P. Bunbury, C. E. Edwards, Proc. Phys. Soc. 77, 1062 (1960)

/7/ S. S. Hanna, L. Meyer-Schützmeister, R. S. Preston, D. H. Vincent, Phys. Rev. 123, 179 (1961)

/8/ G. M. Gorodinskii, L. M. Krizhanskii, E. M. Kruglov, The Magnitude of the Quadrupole Interaction between Sn^{119} Nuclei and a Crystal Lattice /in Russian/ in press.

/9/ B. S. Dzhelepov, L. P. Peker, Decay Schemes of Radioactive Nuclei /in Russian/ Academy of Sciences Press, USSR, 1958

/10/ A.T.F. Boyle, D. St. P. Bunbury, C. E. Edwards, Proc. Phys. Soc. 79, 416, (1962)

/11/ F. L. Shapiro, UFN 72, 685 (1960)

/12/ Yu. Kagan, ZhÉTF 41, 659 (1961)

/13/ Yu. Kagan, ZhÉTF 40, 312 (1961)

/14/ Yu. Kagan, V. A. Maslov, ZhÉTF 41, 1296 (1961)

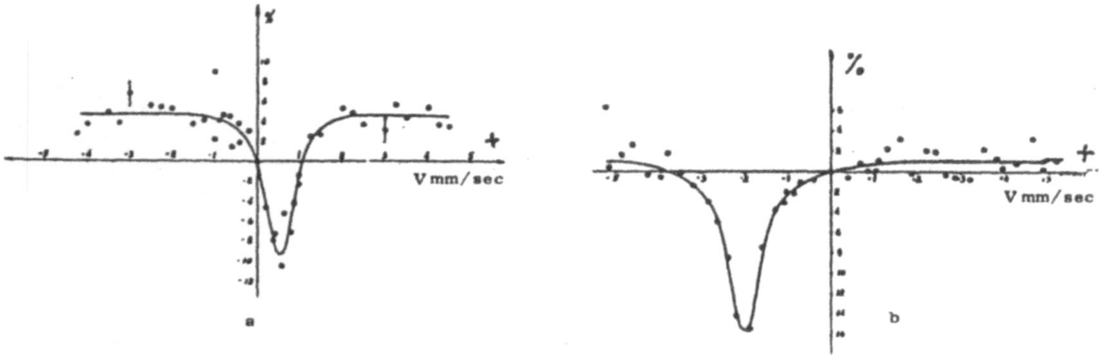

Fig. 1. Resonance absorption spectra in β- Sn (a) and in $Na_2SnO_3 \cdot 3H_2O$ (b). Source α- Sn.

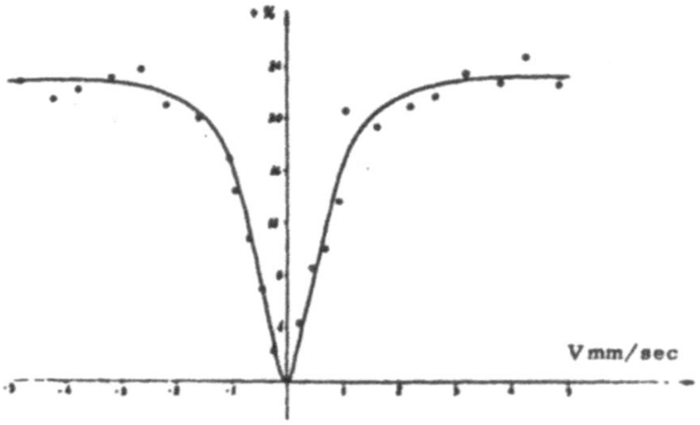

Fig. 2. Resonance absorption spectrum in $Na_2SnO_3 \cdot 3H_2O$. Source SnO_2.

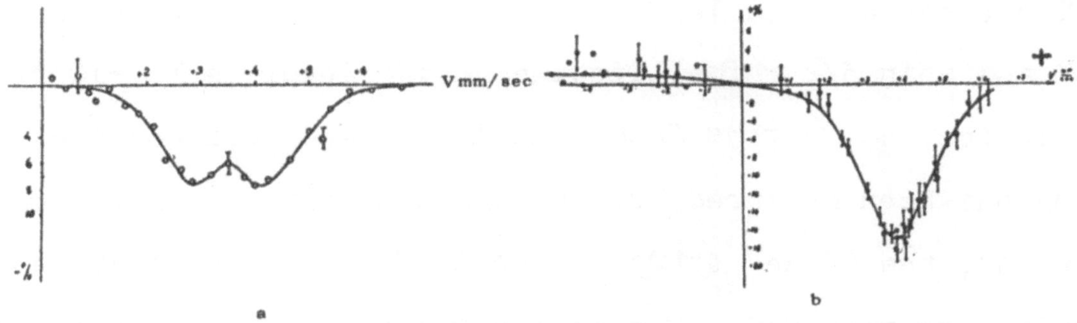

Fig. 3. Resonance absorption curves in $SnCl_2 \cdot 2H_2O$ (a), and in $SnSO_4$ (b). Source SnO_2.

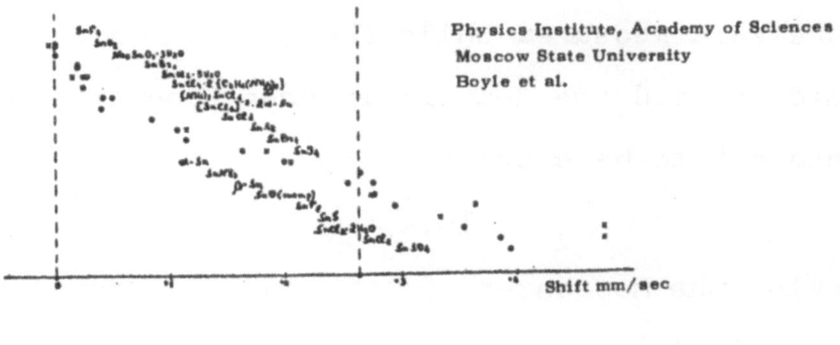

Fig. 4.

OBSERVATION OF THE MÖSSBAUER EFFECT IN Tb^{159}

I. Dezhi, L. Kestkheli

This report gives a short presentation of the preliminary results of measuring the Mössbauer effect in Tb^{159}.

The first excited level of the Tb^{159} nucleus has an energy of 58 keV and spin 5/2. The lifetime of the level is $1.3 \cdot 10^{-10}$ sec /1/. So far, no γ- rays from this level have been observed, and nothing has been measured but the characteristic 44 keV X-rays.

First, the 58 keV γ-transition in Tb^{159} was observed and its intensity was measured. The source was gadolinium nitrate, irradiated in a reactor. After β-decay of the Gd^{159} nucleus with a half life of 18 hours, a soft γ-radiation is emitted which was recorded on a krypton proportional counter. The end of the spectrum (Fig. 1) shows a peak from the 58 keV γ-transition that we were looking for, the peaks at 50 and 44 keV correspond to K_β and K_α of terbium, the peaks at 39 and 32 keV are emergence peaks from K_β and K_α, and the peak at 26 keV corresponds to a γ-transition in Dy^{161}. Several spectra were measured while the gadolinium was decaying, which were used to find the K-shell internal conversion coefficient. It turned out to be equal to

$$\alpha_K = 10.4 \pm 1$$

In observing the Mössbauer effect, the irradiated nitrate was converted to gadolinium oxide, and cooled to liquid nitrogen temperature. The absorber was terbium oxide, Tb_4O_7, with a thickness of 0.1 g/cm^2, fastened to a loud speaker diaphragm. The γ- ray absorption in the terbium oxide was measured. The single channel analyzer slot was set for a 58 keV peak. The background was about 50% of the count in the slot.

The magnitude of the effect $(I_\infty / - I_0)/I_\infty$ was $(1.96 \pm 0.3)\%$.

The result obtained shows that recoilless resonance absorption is found in the 58 keV γ-line of Tb^{159}.

An attempt was made to get some approximate data on the line width. Figure 2 shows how the magnitude of the Mössbauer effect varies with the voltage on the loud speaker winding for iron and terbium. It may be seen from the figure that the line is very broad.

If it is assumed that the lifetime of the level is 10^{-9}--10^{-10} sec, the width of the 58 keV Tb^{159} line will be approximately a factor of 100 greater than the width of the iron line.

A more detailed study of the velocity spectrum will be made later on.

Central Physics Research Institute, Budapest

Translated by: Charles V. Larrick

REFERENCES:

/1/ "Gamma Rays" /in Russian/ Academy of Sciences Press, USSR, 1961 p. 187.

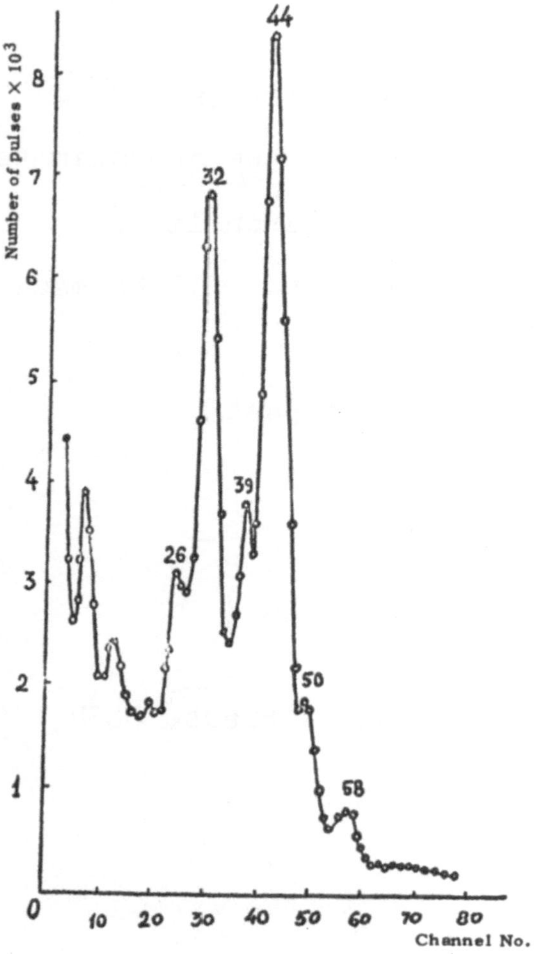

Fig. 1.

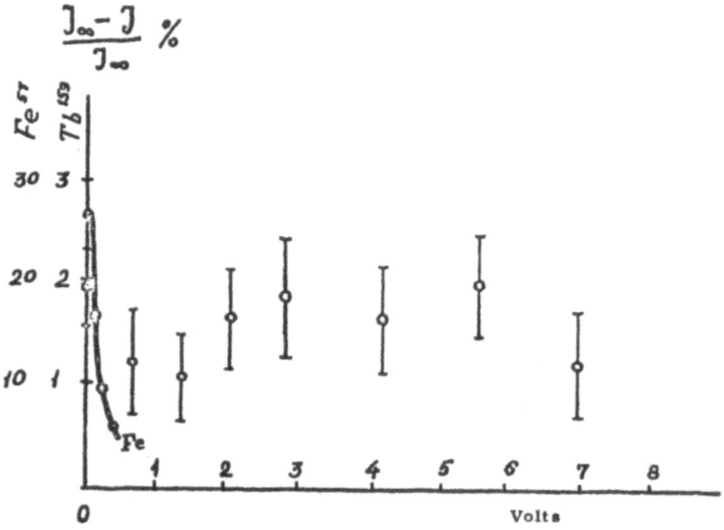

Fig. 2

THEORETICAL INVESTIGATIONS

THE MOSSBAUER EFFECT AND THE THEORY OF RELATIVITY

Ya. Smorodinskii

1. INTRODUCTION

Starting from the autumn of 1959, possibilities of a check of the general theory of relativity by means of the Mossbauer effect have been discussed in the literature /1,2/. The experiments of Pound and others /3,4,5/ have demonstrated that one can observe the influence of a gravitational field on the frequency of a photon under laboratory conditions and that the magnitude of the effect is in agreement with the predictions of the theory. (Compare the discussion of Sherwin /6/). The question arises: what precisely is verified in Pound's experiments? In order to check any theory, we must first of all decide on which features we shall regard as already established and which we question. It is simplest of all when an experiment should distinguish between two theories which give different predictions. In this case we seek an answer in the form of "yes" or "no", and generally no uncertainty arises in the interpretation of the experiments.

In the case of the general theory of relativity, the question turns out to be more complex. There isn't any other theory which can combine the special theory of relativity with a gravitational field. The general theory of relativity is connected with the other branches of physics by such a strong logical chain that any check on it reduces, in the long run, to a check on the conclusions of the special theory of relativity or even more simply of the law of conservation of energy. We shall limit the problem and inquire: what direct conclusions do the results of the experiment of Pound et al allow one to make about the properties of a gravitational field without utilizing other considerations which lead to construction of a theory of gravitation. Let us first recall the principal feature of Schwarzschild's solution which describes the nature of the gravitational field of a material body (see /7,8/ for details).

2. SCHWARZSCHILD'S SOLUTION

The metric of a spherically symmetric field which specifies the Newtonian attraction at large distances is usually written in the form:

$$ds^2 = (c^2 + 2\phi)dt^2 - \frac{dr^2}{1+\frac{2\phi}{c^2}} - r^2 d\Omega^2 \tag{2.1}$$

Here $\phi(r)$ is Newton's potential $\varkappa M/r$ and $r^2\, d\Omega^2$ is an element of area of a sphere. The metric in the form of (2.1) specifies: a) A time coordinate such that the coefficients are time independent; b) A scale chosen so that the area of a sphere is always equal to $4\pi r^2$ while the radius of the sphere will always be less than r. Bear in mind that dr differs from the Euclidian element of length by a quantity of the order of $1/c^2$; setting $\phi \sim gh$ — the height above the surface of the earth, we obtain $\phi/c^2 \sim gh/c^2$. Under laboratory conditions $h \sim 10$ m, $\phi/c^2 \sim 10^{-15}$ *). A quantity of the same order also determines the difference in the rate of clocks at different points.

The coordinate r enters as the argument of the potential $\phi(r)$. If we confine ourselves only to terms of the order of $1/c^2$, it does not matter how we define r and we can use its Euclidian value. If, however, we are interested in higher order effects, then the question of determining distances requires separate consideration.

The space metric can be changed by a transformation of the coordinates. It is convenient to use the so-called isotropic metric, which is obtained

---

*) At the surface of the sun, $\phi/c^2 \sim 2\cdot 10^{-6}$. This number also characterizes the precision of astronomical experiments (compare /8/).

from (2.1) by the substitution

$$r = r_1 \left(1 + r_0/4r_1 \right) , \tag{2.2}$$

where $r_0 = \varkappa M/c^2$ — the gravitational radius of the source, so that $\phi/c^2 = r_0/r$. In terms of the variable r_1 the metric has the form

$$ds^2 = \frac{\left(1 - \frac{r_0}{4r_1}\right)^2}{\left(1 + \frac{r_0}{4r_1}\right)^2} c^2 dt^2 - \left(1 - \frac{r_0}{4r_1}\right)^2 \left[dr_1^2 + r_1^2 d\Omega^2\right]. \tag{2.3}$$

In such a metric, the value of an element of length is independent of direction in accordance with the generally accepted measurement of length with a solid scale. We note that the velocity of light depends on the coordinates in both metrics

$$v_{\text{light}}^2 = c^2\left(1 - \frac{2r_0}{r}\right) = \frac{\left(1 - \frac{r_0}{r_1}\right)^2}{\left(1 + \frac{r_0}{r_1}\right)^4} . \tag{2.4}$$

It is important that the difference in the metric arises only starting with the terms $\sim c^4$.

3. THE FREQUENCY OF A QUANTUM IN A GRAVITATIONAL FIELD

The general conclusion from a frequency measurement in a gravitational field reduces to the following: We proceed from the facts that:

1. The number of oscillations of a quantum between two events does not depend on the observer;

2. The frequency of radiation, or the internal properties of a radiating system, at rest with respect to an observer and at the same location, does not depend on the location or on the observer. This means that we assume that the influence of the gravitational field on the properties of a nucleus or of an atom is negligibly small. More precisely, we assume that Planck's constant is independent of the gravitational field and that the entire dependence is

due only to the dependence of the mass and energy of the system on its coordinates.

Under these assumptions, the product $\omega \Delta t$ will be invariant. Setting $\phi = 0$ and $\omega = \omega_0$ at the surface of the earth, we can write the formula for the frequency corresponding to the potential ϕ:

$$\omega = \omega_0 \left(1 - \frac{2r_0}{r}\right)^{1/2} \tag{3.1}$$

for the metric (2.1), or else in the form

$$\omega = \omega_0 \left(\frac{1 - \frac{r_0}{4r_1}}{1 + \frac{r_0}{4r_1}} \right)^2 \tag{3.2}$$

in the metric (2.2). The difference in the formulas is determined by the different choice of the coordinate r in (3.1). It is convenient to write both formulas in a single form:

$$\omega = \omega_0 \left(1 + \frac{2\phi}{c^2}\right)^{1/2}. \tag{3.3}$$

In the metric (2.1) ϕ coincides with the Newtonian potential $\phi = -r_0/r$; in the isotropic metric

$$\phi = - \frac{\frac{r_0}{r}}{\left(1 + \frac{r_0}{4r_1}\right)^2}, \tag{3.4}$$

which differs from the Newtonian potential at terms of the order of c^{-2}.

The written form of (3.3) is convenient inasmuch as in it the potential is taken to have the character of a natural coordinate, almost equal to the reciprocal distance (in units of r_0).

One can still change the definitions slightly, having written (3.3) in the form

$$\omega = \omega_0 \left(1 + \psi/c^2\right) \tag{3.5}$$

so that

$$\psi = -1 + \sqrt{1 + 2\phi} \tag{3.6}$$

Ψ differs from Φ only by terms of the order of c^{-4}. Since Ψ and Φ are clearly expressed through the coordinates r or r_1, one can introduce Φ^{-1} or Ψ^{-1} in the nature of a coordinate in the metric (2.1) or (2.3) such that the coordinate will be measured directly by the potential.

From what has been said, it is seen that the experiment itself, with a measurement of the frequency shift, assigns the scale of distances. One can give information about the metric of a space only after the measurement of a distance in ordinary units of length. Therein, measurements can be made in the Euclidian approximation to the order of c^{-2}; to go to the order of c^{-4} it is necessary to measure distance to an accuracy inclusive of terms in c^{-2}. A determination of corrections of even higher order already requires taking account of gravitational radiation (of the order of c^{-5}) and the entire problem becomes very complex, even theoretically.

Let us now turn to the experiment with the Mossbauer effect. The experiments of Pound and Rebka, and of Cranshaw, Schiffer and Whitehead, consisted in a comparison of the frequency of a photon emitted by an excited Fe^{57} nucleus at a height of about 10 m above the surface of the earth with the resonance frequency of absorption of a non-excited Fe^{57} nucleus at the surface of the earth. The comparison was carried out such that the photon was absorbed by a target which was moving at just the right velocity to compensate the gravitational frequency change, thanks to the doppler shift.

We shall show that the result of the experiment can be calculated, using only the law of conservation of energy.

At the height h the excited nuclei have a mass $M_1 (1 + gh)$, nuclei in the ground state have a mass $M_o (1 + gh)$. Let a nucleus be at rest before radiating; after radiating it has received a certain recoil velocity U. The nucleus on the ground has a mass M_o before absorption of a quantum, and after absorption — a mass $M_1 = M_o + \Delta M$ *). Let us find what velocity a nucleus on the ground must have in order to be able to absorb radiation from a nucleus situated at a height h. We shall denote this velocity by v, and the increase in velocity after absorption by Δv.

To determine the three velocities U, v and $v + \Delta v$ one must first of all use the laws of conservation of energy and momentum. We obtain a third equation having noted that, upon radiating an electromagnetic wave, the momentum imparted to the nucleus is equal to the imparted energy.

Thus the energy transferred is equal to $\Delta M (1 + gh)$, and the law of conservation of momentum gives (after dividing by M_o):

$$\frac{\Delta M}{M_o}(1 + gh) = \frac{\Delta M}{M_o} v + \Delta v \tag{3.7}$$

(The left side is the momentum of the upstairs nucleus after radiation, the right side is the change in momentum of the downstairs nucleus upon absorption).

The law of conservation of energy gives:

$$\frac{U^2}{2} + \frac{\Delta M}{M_o}(1 + gh) = \frac{\Delta M}{M_o} + \frac{v^2}{2} \cdot \frac{\Delta M}{M} + v \Delta v \tag{3.8}$$

(On the left — loss of energy upstairs, on the right — gain in kinetic energy downstairs).

*) Here it is essential that the nucleus is a quantum object, and that we can speak about the very same two levels of the nucleus, both at ground level and at the height h. It is very difficult to establish such a correspondence with a classical dipole.

Therefore, the quantum nature of the phenomenon turns out to be no less important than the relativistic.

Neglecting the squared velocities, we obtain from (3.7) and (3.8)

$$\Delta \nu = \Delta M / M \qquad (3.9)$$

$$\nu = gh \,. \qquad (3.10)$$

The relation (3.10) was also verified in the experiments of Pound and Rebka.

In the derivation presented, we have used the conservation laws of the special theory of relativity and have considered that the gravitational field has potential character.

In these assumptions there is nothing of the general theory of relativity. The effect of change of energy of a level in the gravitational potential field is analogous to the Stark effect in an electric field which is also determined by the laws of conservation.

The experiments of Pound and Rebka seem analogous to the old experiments of Galileo, who dropped a stone from the tower of Pisa; only in the new experiments mass was carried from the tower to the ground not by a stone, but by an electromagnetic field.

In order that an experiment of the type described should give more complete information about the geometry, it's necessary, at least, to measure the distance between the source and the detector. To do this, for example, one has to measure the time taken by light on the path from the detector to the source and back. This measurement should have a relative precision of the order of ϕc^{-2}, i.e. of the order of 10^{-15}. In this case we would obtain information about the spatial part of the metric.

What has been said can be illustrated by a geometrical example. The measurement of the frequency shift is simply an establishment of a scale along the axis of distances or along the time axis, but there is no establishment of a correspondence between the scales of the two axes. It is obvious that one can not determine the curvature of space-time through measurements along the coordinate axes. Measurement of the time of propagation of light can be described as a measurement of the base of an isosceles triangle Δt in the (x,t) plane. In this triangle, furthermore, one knows the height —the distance from the source and the the angles at the base, determined from the velocity of light. Knowing four elements of a triangle (two angles, the altitude and the base) one can find by how much it differs from a triangle in Euclidian geometry and calculate the curvature.

It is clear that the frequently discussed experiments (compare /9/) on the frequency shift at an artificial satellite do not give more information than do laboratory experiments, since in this case one has to have a method to measure the time of propagation of a signal.

Joint Institute of Nuclear Studies

REFERENCES

/1/ I.Ya. Barit, M.I. Podgoretskii and F.L. Shapiro, ZhETF, 38, 301 (1960).

/2/ R.V. Pound and G.A. Rebka, Jr. Phys. Rev. Lett. 3, 439 (1959).

/3/ R.V. Pound and G.A. Rebka, Jr. Phys. Rev. Lett. 4, 337 (1960).

/4/ T.E. Cranshaw, J.P. Schiffer and A.B. Whitehead, Phys. Rev. Lett. 4, 163 (1960).

/5/ H.J. Hay, J.P. Schiffer, T.E. Cranshaw and P.A. Egelstaff, Phys. Rev. Lett. 4, 165 (1960).

/6/ C.W. Sherwin, Phys. Rev. 120, 17 (1960).

/7/ A. Eddington, Theory of Relativity, /Russian translation/ Fizmatgiz, M-L 1934.

/8/ L. Landau and E. Lifshits, Field Theory, 3rd ed. /in Russian/ GTTI M-L 1962.

/9/ V.L. Ginzburg, in the symposium "Einstein and the development of physical-mathematical thought." Izd. Akad. Nauk SSSR 1962, pg. 117.

Translated by Robert L. Eisner

EMISSION AND SCATTERING OF γ-RAY QUANTA BY THE NUCLEI IN A SOLID IN THE PRESENCE OF A TIME-DEPENDENT EXTERNAL PERTURBATION

I. P. Dzyub and A. F. Lubchenko

ANNOTATION

A theory is developed of the emission and scattering of γ-ray quanta by the nuclei in a solid in the presence of a time-dependent external perturbation. The shapes of the emission and scattering spectra are calculated on the assumption that the time dependence of the perturbation is periodic. It is shown that the presence of such a perturbation leads to the appearance in the spectrum of additional Mössbauer lines in addition to the main one; in the spectrum of the scattering the main line appears in observations made at the Bragg angle; the angle at which the additional lines are observed depends on the wave vector of the lattice vibration produced by the external perturbation. This makes it possible to determine the frequency spectrum of the normal vibrations of a solid from the spectrum of the scattered quanta.

1. INTRODUCTION

The study of the Mössbauer spectra /1/ of the emission, absorption, and scattering of γ-ray quanta by the nuclei in a solid makes it possible to obtain definite information not only about the γ-radioactive nucleus involved in the γ-ray transitions, but also about the solid which has the nucleus as a constituent. By this method values have been found for the internal magnetic field in ferromagnetics /2-4/, the self-diffusion coefficient of atoms /5/, and the index of refraction of a solid for γ-rays with energies of some tens of keV /6/. Another very interesting use of the Mössbauer effect is the determination of the frequency spectrum w(f) of the normal vibrations of a solid. In /7-8/ the spectra of the resonance and Rayleigh scattering of γ-ray quanta by the nuclei in a solid have been studied, and it has been shown that in principle it is possible to use the Mössbauer effect in no-phonon and one-phonon γ-ray transitions to determine w(f). In practice, however, the use of these spectra to determine w(f) is complicated by the fact that the intensity of the one-phonon line is smaller than that of the no-phonon Mössbauer line by more than an order of magnitude and it is extremely difficult to detect such lines experimentally.

A question which naturally arises is that of strengthening particular normal vibrations, for example by the generation of ultrasonic vibrations in the solid /9/ which serves as the scatterer, and using the spectrum of the scattered γ-rays from such a scatterer to determine w(f).

In this paper we calculate the spectrum of the γ-rays scattered by a solid in which, in addition to their thermal vibrations, the atoms are executing forced vibrations caused by an external action. We estimate the intensities of the no-phonon and one-phonon Mössbauer lines and show that they can differ by less than an order of magnitude, and thus can be of practical use for the determination of the frequency spectrum of the normal vibrations of a solid.

1. EMISSION OF γ-RAY QUANTA BY NUCLEI IN A SOLID IN THE PRESENCE OF AN EXTERNAL PERTURBATION

We consider a system consisting of a crystal containing γ-radioactive nuclei, which in addition to their thermal vibrations are executing forced vibrations under the action of a time-dependent external perturbation, and the radiation field. The Hamiltonian of the system is

$$H_0 + H' + H''$$

where H' is the energy of the interaction between the radiation field and the crystal, and $H'' = \sum_{jn} F_{jn}(t)\, u_{jn}$ is the operator for the energy of the external perturbation; $F_{jn}(t)$ is the external force acting on the <u>j</u>th atom of the <u>n</u>th cell, and u_{jn} is the displacement of this atom from equilibrium;

$$H_0 = H_c + H_\sigma,$$

where H_σ is the operator for the energy of the electromagnetic field, and H_c is the operator for the energy of the crystal, in which the vibrations of the nuclei are treated in the harmonic approximation. The wave functions and corresponding eigenvalues of the Hamiltonian H_0 are

$$\psi_{\ell(n_s)(N_\sigma)} = \varphi_\ell\, |n_s\rangle |N_\sigma\rangle,$$

$$E_{\ell(n_s)(N_\sigma)} = \mathcal{E}_\ell + \sum_s \hbar\omega_s n_s + \sum_\sigma \hbar\nu_\sigma N_\sigma$$

Here φ_ℓ is the wave function which describes the state of the nucleus and the electrons of the crystal (ℓ is a complete set of quantum numbers for the state), $|n_s\rangle$ and $|N_\sigma\rangle$ are respectively the wave functions for the phonon and photon fields in the occupation-number representation, ω_s are the frequencies of the normal vibrations of the solid, the index s numbers the values of the wave vector f and the branch τ of the vibrations of the crystal lattice, ν_σ are the frequencies of the electromagnetic field, and ε_l is the energy of the crystal when the phonon-field occupation numbers are equal to zero.

To calculate the probability of the transition

$$l_1 (n_s)(0) \longrightarrow l_0 (n'_s)(1\sigma),$$

in which the nucleus and electrons, in the presence of the perturbation H'', go from an excited state l_1 to the ground state l_0 with the emission of a γ-ray quantum of frequency ν_σ, it is necessary to calculate the matrix element of the S matrix

$$\langle l_0 n' t_\sigma | T \exp\left\{-\frac{i}{\hbar}\int_{-\infty}^{+\infty}[H'(t) + H''(t)]\, dt\right\} | l_1 n_s 0 \rangle,$$

where $H'(t)$ and $H''(t)$ are the operators H' and H'' in the interaction representation and T is the chronological operator. When we here expand $\exp\{-iH'(t)/\hbar\}$ in a series of powers of $H'(t)$ and confine ourselves to the first term of the expansion, we get as the expression for the transition probability

$$W_{n_s} = \frac{1}{\hbar^2} \left| \langle l_0 n'_s t_\sigma | T \int_{-\infty}^{+\infty} dt\, H'(t) S_1 | l_1 n_s 0 \rangle \right|^2,$$

where

$$S_1 = T \exp\left\{-\frac{i}{\hbar}\int_{-\infty}^{+\infty} H''(t)\, dt\right\}. \qquad (1)$$

If we neglect the reabsorption of the γ-rays by the electrons and nuclei in the crystal, and use the fact that in nuclear phototransitions there is very little change of the electronic state, we can write the operator H' in the form

$$H' = \sum_{\sigma} L^{j}_{\sigma} e^{i\varkappa_\sigma R_{nj}} (a_\sigma + a^{+}_{-\sigma}) + CC, \qquad (2)$$

where $R_{nj} = R^0_{nj} + u_{nj}$, $R^0_{nj} = R^0_n + r_j$ gives the radius vector of the equilibrium position of the (n_j)th nucleus in the lattice, $\varkappa$ is the wave vector of the γ-ray quantum, L^j_σ is an operator which depends on the coordinates of the nucleons, and a^+_σ, a_σ are operators for creation and destruction of quanta of the electromagnetic field. The expression for W_{n_s} can now be written in the form

$$W_{n_s} = \frac{|L^j_{01}|^2}{\hbar^2} \left| \langle n'_s | T \int_{-\infty}^{+\infty} e^{i(\Omega_{l_1 l_0} - \nu_\sigma)t - i\varkappa_\sigma u_{nj}(t)} S_1 | n_s \rangle \right|^2. \qquad (3)$$

Here

$$L^j_{01} = \langle l_0 | L^j_\sigma | l_1 \rangle, \quad \Omega_{l_1 l_0} = \frac{\mathcal{E}_{l_1} - \mathcal{E}_{l_0}}{\hbar}$$

$$u_{nj}(t) = \exp\left\{\frac{iH_f t}{\hbar}\right\} u_{nj} \exp\left\{-\frac{iH_f t}{\hbar}\right\}, \qquad (4)$$

and $H_f = \sum_s \hbar\omega_s b^+_s b_s$ is the operator which describes the vibrations of the nuclei in the crystal.

To get the total transition probability we must sum the quantity (3) over all final states of the crystal lattice (n'_s) and average over all initial states (n_s); when we also carry out the chromological ordering, we can write the expression for the transition probability in the form*

$$W_{\infty} = \frac{|L_{01}|^2}{\hbar^2} \int_{-\infty}^{+\infty} dt_2 \int_{-\infty}^{+\infty} dt_1 \, e^{i(\Omega_{l_1 l_0} - \nu_\sigma)(t_2 - t_1)} \langle T_c \exp\left(-\frac{i}{\hbar} \int_C V(t)\,dt\right) \rangle, \qquad (5)$$

where $V(t) = H''(t) + \hbar \varkappa_\sigma u(t) [\delta(t - t_1) - \delta(t - t_2)]$, $\langle \ldots \rangle = \frac{1}{z} Sp\{e^{-\beta H_f} \ldots\}$, $\beta = \frac{1}{KT}$, z is the partition function for the phonons, and the path of integration c is shown in Fig. 1.

---

*Here we omit the indices n and j which number the cells and the nuclei in a cell.

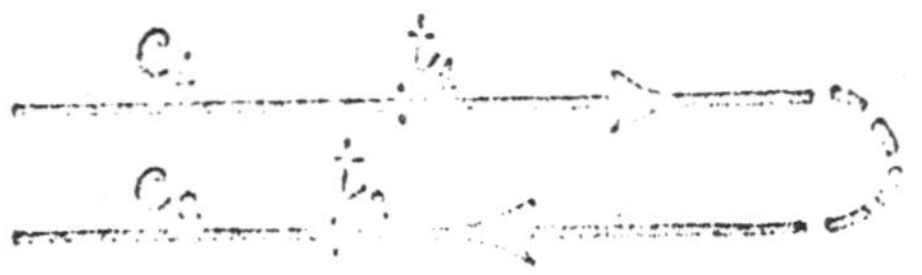

Fig. 1

To calculate the expression $\langle T_c \exp\{-\frac{i}{\hbar}\int_c V(t)dt\}\rangle$ we expand the exponential in a series of powers of $-\frac{i}{\hbar}\int V(t)\,dt$. Then, using the fact that

$$u_{nj} = \sum \left(\frac{\hbar}{2M_j N\omega_s}\right)^{1/2} V_j(f,\sigma)\, e^{ifR_n^0}\left(b_{fs} + b^{+}_{-fs}\right) \tag{6}$$

where M_j is the mass of the jth atom, N is the number of atoms in the crystal, and $V_j(f, \tau)$ is the displacement of the jth atom in the cell n = 0 /10/, we can easily show that

$$\langle T\, V(t_1) V(t_2) \ldots\ldots V(t_{2n+1})\rangle = 0$$

To calculate the terms which contain an even number of operators V(t), we use a form of Wick's theorem generalized to the case of temperatures different from zero /11-12/. Then we can represent the integrand of each term of the series in the form of a sum of products of average values of pairings, of the type

$$\langle T\, V(t_n)\, V(t_m)\rangle$$

Term number 2n is then of the form

$$\left(-\frac{1}{2\hbar^2}\right)^n \frac{1}{n!}\left\{\int_c dt \int_c dt' \langle T\, V(t) V(t')\rangle\right\}^n$$

Summing the series so obtained we find that

$$\langle T_c \exp\left(-\frac{i}{\hbar}\int_c V(t)\,dt\right)\rangle = \exp\left\{-\frac{1}{2\hbar^2}\int_c dt \int_c dt' \langle T\, V(t) V(t')\rangle\right\}$$

This last integral can be most simply calculated by breaking the path c up into parts C_1 and C_2 and then, carrying out the chronological ordering, we get

$$\frac{1}{2\hbar^2}\int_c dt \int_{c'} dt' \langle T\, V(t) V(t')\rangle = C(T) - i\varkappa_\sigma \bar{u}(t_1) + i\varkappa_\sigma \bar{u}(t_2) - \sum_{\alpha,\alpha'=1}^{3} \varkappa_\sigma^{\alpha}\varkappa_\sigma^{\alpha'}\left\{\frac{\langle u^{\alpha}(t_1)u^{\alpha'}(t_1)\rangle + \langle u^{\alpha}(t_2)u^{\alpha'}(t_2)\rangle}{2} - \langle u^{\alpha}(t_2)u^{\alpha'}(t_1)\rangle\right\}, \tag{7}$$

where

$$\bar{u}^{\alpha'}(t) = \frac{1}{i\hbar}\sum_{nj\alpha}\int_{-\infty}^{+\infty} dt_1\, F_n^{\alpha}(t_1)\,\langle\left[u^{\alpha'}(t), u^{\alpha}_{nj}(t_1)\right]\rangle$$

is the average displacement of the atom produced by the force $F_{jn}(t)$, and $c(\tau)$ is a sum of expressions of the type

$$\sum_{\substack{nj\alpha \\ n'j'\alpha'}} \int_{C_1}\int_{C_2} dt_1\, dt'\, F^{\alpha}_{jn}(t) F^{\alpha}_{j'n'}(t) < u^{\alpha'}(t),\, u^{\alpha}_{nj}(t_1) >$$

It is easily shown that $b_s(t) = b_s \exp(-i\omega_s t)$, and therefore

$$\langle u^{\alpha}(t_2) u^{\alpha'}(t_1)\rangle = \sum_s \bar{q}^{\alpha}(s) \bar{q}^{\alpha'}(-s)(2\bar{n}_s + 1),$$
$$\langle u^{\alpha}(t_2) u^{\alpha'}(t_1)\rangle = \sum_s \bar{q}^{\alpha}(s) \bar{q}^{\alpha'}(-s)\left\{\bar{n}_s e^{i\omega_s(t_2-t_1)} + (\bar{n}_s+1) e^{-i\omega_s(t_2-t_1)}\right\}, \quad (8)$$

where $\bar{n}_s = \left(e^{\frac{\hbar\omega_s}{KT}} - 1\right)^{-1}$ is the Planck average for the phonons, and

$$\bar{q}^{\alpha}(s) = \frac{V_j^{\alpha}(s) \exp\{i f R^{\circ}_n\}}{(2M_j N \hbar^{-1} \omega_s)^{1/2}} \quad u \quad \bar{q}^{\alpha}(s) = \bar{q}^{*\alpha}(s) . \quad (9)$$

When we now set $\sum \bar{q}^{\alpha}(s) \varkappa^{\alpha}_{\sigma} = p^{s}_{\sigma j}$, we can write the probability for a transition with the emission of a γ-ray quantum in the form

$$W_{\alpha} = \frac{|L'_{01}|^2}{\hbar^2} \int_{-\infty}^{\infty} dt_2 \int_{-\infty}^{\infty} dt_1\, e^{i(\Omega_{\ell_1\ell_0} - \nu_\gamma)(t_2-t_1) - i\varkappa_\sigma \bar{u}_{nj}(t_1) + i\varkappa_\sigma \bar{u}_{nj}(t_2) + g(t_2-t_1)}, \quad (10)$$

where

$$L'_{01} = L_{01} \exp\{-\tfrac{1}{2} c(T)\},$$
$$g(t_2-t_1) = \sum_s |p^{s}_{\sigma j}|^2 \left\{\bar{n}_s e^{i\omega_s(t_2-t_1)} + (\bar{n}_s+1) e^{-i\omega_s(t_2-t_1)} - (2\bar{n}_s+1)\right\} . \quad (11)$$

We note that the expression (10) takes the perturbation acting on the atoms of the crystal into account exactly; it does not depend on the form of the function $F_{jn}(t)$. For the concrete calculation of the shape of the emission spectrum we shall hereafter assume that the perturbation is periodic and set

$$\bar{u}_{nj}(t) = a_j \cos(\omega_q t - q R^{\circ}_n),$$

where ω_q and a_j are respectively the frequency and the amplitude of the forced vibrations of the atoms of the crystal, and q is the wave vector of the vibration. Then, using the relation

$$e^{ia\cos\varphi} = \sum_{K=-\infty}^{\infty} i^{K} J_K(a) e^{iK\varphi}, \quad (12)$$

where $\mathcal{J}_\kappa(a)$ is the Bessel function of order κ, we can represent the probability per second for the emission of a γ-ray quantum /13/ in the form

$$W_1 = \frac{2|L'_{01}|^2}{\hbar^2} \sum_{\kappa=-\infty}^{\infty} \mathcal{J}_\kappa^2(\varkappa a_j)\, \mathrm{Re} \int_0^\infty dx\, e^{i(\Omega_{4,1_0} - \nu_0 + \kappa\omega_g)x + g(x)}, \quad (13)$$

where $g(\kappa)$ is defined by Eq. (11).

It follows from this that the emission spectrum will consist of a phonon part arising from the thermal vibrations of the atoms in the crystal, with clearly distinguishable Mössbauer lines: the no-phonon line corresponding to $\kappa = 0$ in Eq. (13), one-phonon lines corresponding to transitions without change of the occupation numbers of the thermal phonon field, but with a change of $\pm\hbar\omega_g$ in the energy of the forced vibration of the emitting atom $[\kappa = \pm 1$ in Eq. (13)$]$. If the amplitude of the forced vibrations is such that $\varkappa a_j = 0.7$, the intensities of the corresponding lines will be proportional to 0.77 and 0.11; the two-phonon lines corresponding to $\kappa = \pm 2$ in Eq. (13) will be practically invisible in the spectrum.

We note that a similar shape of the emission spectrum is obtained when we take the Doppler effect into account, with the emitter moving as a whole /14/.

The results obtained above show that the Mössbauer emission spectra can be used in particular for the determination of the vibration frequencies of ultrasonic generators.

2. RAYLEIGH SCATTERING OF γ-RAY QUANTA BY THE NUCLEI IN A SOLID IN THE PRESENCE OF AN EXTERNAL PERTURBATION

To determine the shape of the spectrum of γ-rays scattered by a scatterer which is subject to a time-dependent external perturbation, we must calculate the martix element of the S matrix in second order in the electromagnetic interaction

$$S^{(2)}_{l_0 l_0}(n_j) = \langle l_0 n'_s, N_{\sigma'}-1, N_{\sigma''}+1 \left| \frac{1}{\hbar^2} \int_{-\infty}^{+\infty} dt_1 \int_{-\infty}^{+\infty} dt_2 \, H'(t_1) H'(t_2) S_1 \right| l_0 n_s^0 N_\sigma \rangle ,$$

where $H'(t)$ and S_1 are defined as in the case of emission, and n_s^0 and n'_s are the occupation numbers of the phonon field in the initial and intermediate states of the scatterer.

As in the case of emission, we neglect the change of the electronic state of the crystal in nuclear phototransitions, and take the operator H' in the form (2). Then, denoting the state of the nucleus and electrons in the intermediate state by l (l'), we find for the matrix $S^{(2)}_{l_0 l_0} = \sum_{n_j} S^{(2)}_{l_0 l_0}(n_j)$ the expression

$$S_{l_0 l_0} = \frac{\sqrt{N_{\sigma'}(N_{\sigma''}+1)}}{\hbar^2} \sum_{n_j} \sum_{l} L^{j}_{l_0 l}(\nu_{\sigma''}) L^{j}_{l l_0}(\nu_{\sigma'}) \langle n'_s \left| \int_{-\infty}^{+\infty} dt_1 \int_{-\infty}^{+\infty} dt_2 \exp\{ it_1(\Omega_{l_0 l} + \nu_{\sigma''}) + \right.$$

$$\left. + it_2(\Omega_{l l_0} - \nu_{\sigma'})\} \exp\{-i\varkappa_{\sigma''} R_{nj}(t_1)\} \exp\{ i\varkappa_{\sigma'} R_{nj}(t_2)\} S_1 \right| n_s^0 \rangle , \qquad (14)$$

where $L^{j}_{l l_0}(\nu_{\sigma''}) = \langle l_0 | L^{j}_{\sigma''} | l \rangle$, and in obtaining Eq. (14) we have neglected terms which in the case $u_{nj} = 0$, $H'' = 0$ are proportional to $(\Omega_{l l_0} - \nu_{\sigma''})^{-1}$. To get the total probability W_∞ for scattering of a γ-ray quantum by a crystal in the presence of a time-dependent external perturbation, we must take the square of the absolute value of $S^{(2)}_{l_0 l_0}$ and then sum the resulting expression over all numbers of phonons in the final state and average over all sets of n_s^0 in the initial state. Then

$$W_\infty = \sum_{(n_s)(n_s^0)} W(n_s^0) \left| S^{(2)}_{l_0 l_0} \right|^2 ,$$

where $W(n_s^0) = \frac{1}{z} \exp\{-\beta \sum_s \hbar\omega_s n_s^0\}$, $\beta = 1/kT$ and z is the partition function for the phonons. When we then carry out the chronological ordering of the operators in this expression, we get as the expression for W_∞

$$W_\infty = \frac{\overline{N_{\sigma'}}(\overline{N_{\sigma''}}+1)}{\hbar^4}\sum_{l,l'}\sum_{\substack{nj\\n'j'}} L^{j}_{l,l}(\nu_{\sigma''},\nu_{\sigma'}) L^{j'}_{l'l_0}(\nu_{\sigma''},\nu_{\sigma'}) e^{i\varkappa\Delta R} A_{nj,n'j'}(l,l'), \quad (15)$$

where

$$\varkappa = \varkappa_{\sigma'} - \varkappa_{\sigma'}, \quad \Delta R = R^{\circ}_{n'j'} - R^{\circ}_{nj}, \quad L^{j}_{l_0 l_0}(\nu_{\sigma''},\nu_{\sigma}) = L^{j}_{l_0 l}(\nu_{\sigma''}) L^{j}_{l l_0}(\nu_{\sigma'}),$$

$$A_{nj,n'j'}(l,l') = \int_{-\infty}^{\infty} dt_3 \int_{-\infty}^{t_3} dt_4 \int_{-\infty}^{\infty} dt_1 \int_{-\infty}^{t_1} dt_2 < T_c \exp\left\{-\frac{i}{\hbar}\int_c V(t)\,dt\right\} \times$$

$$\times \exp\{it_1(\Omega_{l_0 l} + \nu_{\sigma''}) + it_2(\Omega_{l l_0} - \nu_{\sigma'}) - it_3(\Omega_{l_0 l'} + \nu_{\sigma''}) - it_4(\Omega_{l' l_0} - \nu_{\sigma'})\}, \quad (16)$$

$$V(t) = H''(t) + \hbar\{\varkappa_{\sigma''} u_{nj}(t)\delta(t-t_1) - \varkappa_{\sigma'} u_{nj}(t)\delta(t-t_2) + \varkappa_{\sigma'} u_{n'j'}(t)\Delta(t-t_4) - \varkappa_{\sigma''} u_{n'j'}(t)\Delta(t-t_3)\},$$

and $\Delta(t - a)$ has the property that

$$\int_{-\infty}^{+\infty} f(t)\Delta(t-a)\,dt = f(a),$$

and $u_{nj}(t)$ is defined by the expressions (4) and (6).

Just as in the case of emission we find

$$< T_c \exp\left\{-\frac{i}{\hbar}\int_c V(t)\,dt\right\} > = \exp\left\{-\frac{1}{2\hbar^2}\int_c dt \int_c dt' < T\, V(t) V(t') >\right\}.$$

When we now introduce the notation

$$G_j = \sum_s (\bar{n}_s + \tfrac{1}{2})(|\rho^{s}_{\sigma'' j}|^2 + |\rho^{s}_{\sigma' j}|^2),$$

after integrating the last expression over dt and dt' we get

$$< T_c \exp\left\{-\frac{i}{\hbar}\int_c V(t)\,dt\right\} > = \exp\{-G_j - G_{j'} + i\varkappa_{\sigma'}\bar{u}_{nj}(t_2) - i\varkappa_{\sigma'}\bar{u}_{n'j'}(t_4) +$$

$$+ i\varkappa_{\sigma''}\bar{u}_{n'j'}(t_3) - i\varkappa_{\sigma''}\bar{u}_{nj}(t_1) + B_0 + B_1(t_1 t_2 t_3 t_4) - C(\tau)\}, \quad (17)$$

where
$$B_0 = -\sum_{\alpha\alpha'} \varkappa^{\alpha}_{\sigma'} \varkappa^{\alpha'}_{\sigma''} < u^{\alpha}_{n'j'}(t_4) u^{\alpha'}_{nj}(t_2) > + \sum_{\alpha\alpha'} \varkappa^{\alpha}_{\sigma''} \varkappa^{\alpha'}_{\sigma''} < u^{\alpha}_{nj}(t_3) u^{\alpha'}_{nj}(t_1) > +$$

$$+ \sum_{\alpha\alpha'} \varkappa^{\alpha}_{\sigma'} \varkappa^{\alpha'}_{\sigma'} < u^{\alpha}_{n'j'}(t_4) u^{\alpha'}_{nj}(t_2) > - \sum_{\alpha\alpha'} \varkappa^{\alpha}_{\sigma'} \varkappa^{\alpha'}_{\sigma'} < u^{\alpha}_{n'j'}(t_3) u^{\alpha'}_{nj}(t_2) >, \quad (18)$$

$$B_1(t_1 t_2 t_3 t_4) = \sum_{\alpha\alpha'} \varkappa^{\alpha}_{\sigma''} \varkappa^{\alpha'}_{\sigma'} < u^{\alpha}_{nj}(t_1) u^{\alpha'}_{nj}(t_2) > + \sum_{\alpha\alpha'} \varkappa^{\alpha}_{\sigma'} \varkappa^{\alpha'}_{\sigma''} < u^{\alpha}_{n'j'}(t_4) u^{\alpha'}_{n'j'}(t_3) >,$$

and $< u^{\alpha}_{n'j'}(t) u^{\alpha'}_{nj}(t) >$ and $\bar{u}_{nj}$ are respectively defined by Eqs. (7) and (8).

Thus the calculation of W_∞ reduces to integration of the expression (16) after substituting in it the expression (17).

It is easy to see that neglect of the dependence of $B_1(t_1,t_2,t_3,t_4)$ on the times $t_1 - t_2$ and $t_4 - t_3$ is equivalent to neglect of ω_s in comparison with $\Omega_{ll_0} - \nu_\sigma$, which is always permissible far from the region of absorption. Therefore, confining ourselves to such a region, we have

$$-G_j - G_{j'} + B_1(0,0,0,0) = -M_j - M_{j'},$$

where

$$M_j = \sum_s \left(\bar{n}_s + \frac{1}{2}\right) \left| p^s_{\sigma'' j} - p^s_{\sigma' j} \right|^2 .$$

If we now assume that there is no external perturbation, the expressions (15)--(18) will determine the probability of Rayleigh scattering of γ-ray quanta by the nuclei of the solid. In particular, for $\varkappa u_{nj} < 1$ the probabilities per unit time for no-phonon and one phonon transitions can be written in the forms

$$W_1' = \bar{N}_{\sigma'}(\bar{N}_{\sigma''}+1) \left| \sum_n e^{i\varkappa R_n^0} \right|^2 \left| \sum_j c_j e^{-M_j - i\varkappa r_j} \right|^2 \delta(\nu_{\sigma''} - \nu_{\sigma'}) ,$$

$$W_2' = \bar{N}_{\sigma'}(\bar{N}_{\sigma''}+1) \left| \sum_{f,n} e^{i(\varkappa - f) R_n^0} \right|^2 \left| \sum_j c_j x_j^s e^{-M_j - i\varkappa r_j} \right|^2 \times$$

$$\times \left\{ (\bar{n}_s + 1)\delta(\nu_{\sigma''} - \nu_{\sigma'} + \omega_s) + \bar{n}_s \delta(\nu_{\sigma''} - \nu_{\sigma'} - \omega_s) \right\},$$

where c_j is the atomic-nuclear form-factor $x_j^s = \dfrac{(\varkappa V_j(s))}{\sqrt{2 m_j N \hbar \omega_s}}$, $\varkappa = \varkappa_{\sigma''} - \varkappa_{\sigma'}$; and the radius vector r_j fixes the position of the jth atom in the cell. These expressions agree with those obtained in /7/.*

From Eqs. (15)--(18) we can easily obtain the Mössbauer spectrum of the scattered γ-rays, which corresponds to γ-transitions without change of the occupation numbers of the thermal phonon field, if we assume that

$$u_{nj}(t) = a_j \cos(\omega_f t - f R_n^0) .$$

---

*There is an error in Eqs. (9c), (11c), and (17c) of /7/; in these equations $(f r_j)$ must be omitted in the exponents.

In fact, when in Eq. (17) we neglect the quantity B_0, which causes the thermal-phonon part of the scattering spectrum, and use Eq. (12), we can write the scattering probability per unit time in the form

$$W_{1m} = \bar{N}_{\sigma'}(\bar{N}_{\sigma''}+1)\sum_m \delta(\nu_{\sigma''}-\nu_{\sigma'}+m\omega_q)\left|\sum_n e^{-i(\varkappa+mq)R_n^0}\right|^2 \times$$
$$\times\left|\sum_j c_j e^{-M_j - i\varkappa\tau_j} J_m(\varkappa a_j)\right|^2$$

It follows from this that when the scatterer is irradiated with a quasi-monochromatic beam of γ-rays of frequency Ω and observed at the Bragg angle, which is determined from the condition $K_{\sigma''} - K_{\sigma'} = K$, where K is a reciprocal-lattice vector, there will appear in the spectrum of the scattered γ-rays a sharp Mössbauer line of frequency Ω and of intensity proportional to

$$\bar{N}_\Omega(\bar{N}_{\sigma''}+1)\left|\sum_j c_j e^{-M_j - i\varkappa\tau_j} J_0(a_j K)\right|^2;$$

when observations are made at angles given by the expression

$$\varkappa_{\sigma''} - \varkappa_\Omega \pm q = K$$

lines of frequency $\nu_{\sigma''} = \Omega \pm \omega_q$ and intensity proportional to

$$\bar{N}_\Omega(\bar{N}_{\sigma''}+1)\left|\sum c_j e^{-M_j - i\varkappa\tau_j} J_1[a_j(K \pm q)]\right|^2.$$

will appear in the spectrum of the scattered rays. Similar expressions can also be obtained for lines of the frequencies $\nu_{\sigma'} = \Omega \pm m\omega_q$, m = 2, 3 ... . It is easy to see that for a given scatterer at a given temperature the intensities of the lines in the spectrum of the scattered rays are actually determined by the ratio of the amplitude a_j of the forced vibrations to the lattice constant d. If $\left|4\pi \frac{a_j}{d}\right| = 0{,}6$ the lines corresponding to $m = \pm 1$ in the expression (10) will have intensities smaller by an order of magnitude than that of the line with $m = 0$. The intensities of lines with $|m| > 1$ will be much smaller.

By experimentally determining the angle at which the lines

with $m = \pm 1$ are observed, by the usual construction in the reciprocal-lattice space/7/, one can easily determine the corresponding values of q, and thus obtain for various ω_q the frequency spectrum of the normal vibrations of the scatterer.

In conclusion we express our gratitude to N. N. Bogolyubov for his interest in this work and for valuable comments.

Institute of Physics, Academy of Sciences, Ukrainian SSR

REFERENCES

1. R. L. Mössbauer, Zs. f. Physik, 151, 124, 1958; Zs. Naturforsch, 14a, 211, 1959.
2. S. S. Hanna, J. Heberle, C. Littlejohn, ..., Phys. Rev. Lett. 4, 177, 1960.
3. G. K. Wertheim, Phys. Rev. Lett. 4, 403, 1960.
4. D. E. Nagl, H. Frauenfelder, R. D. Taylor, D. R. F. Cochran, B. T. Matthias, Phys. Rev. Lett. 5, 364, 1960.
5. A. J. Boyle, D. St. P. Bunbary, C. Edwards, H. E. Hall, Proc. Phys. Soc. 77, 129, 1961.
6. L. Grodzins, E. A. Phillips, Phys. Rev., 124, 774, 1961.
7. I. P. Dzyub, A. F. Lubchenko, FTT, 3, 2275, 1961.
8. G. Tzara, J. Phys. radium 22, 303, 1961.
9. L. Bergman, Ultrasound and Its Application in Science and Technology (Russian Translation), I. L., Moscow, 1958.
10. R. Peierls, Quantum Theory of Solids, Oxford, 1955.
11. T. Matsubara, Progr. Theor. Phys. 14, 351, 1955.
12. D. J. Thouless, Phys. Rev. 107, 1162, 1957.
13. P. Matthews, The Relativistic Quantum Theory of Elementary

Particle Interactions (Russian Translation), I. L., Moscow, 1959.

14. S. L. Raby, D. I. Bolef, Phys. Rev. Lett., 5, 5, 1960.

Translated by W. H. Furry

THE MÖSSBAUER EFFECT AT IMPURITY NUCLEI

I. P. Dzyub and A. F. Lubchenko

INTRODUCTION

Resonance emission and absorption of γ- quanta by impurity nuclei in a crystal has characteristic properties of its own. This is due to the fact that there is a difference in mass between the impurity atom and the atoms in the ideal lattice, as well as the fact that the binding between the impurity atom and the atoms in the ideal lattice is different from the binding between the atoms normally in the lattice, and this causes the vibrations of the impurity atom to be substantially different in some cases from the vibrations of the atoms in the ideal lattice. For example, local frequencies can appear in the frequency spectrum of the normal vibrations of an ideal lattice, where the amplitude of vibration of the impurity atom at these frequencies is greatly different from the amplitude of the vibrations of the atoms in the ideal lattice /1--5/. And since the amplitude of the Mössbauer effect depends on the amplitude of the vibration of the absorbing (or emitting) nucleus, it becomes obvious that there is a relation between the intensity of the Mössbauer line and the nature of the nuclear vibration. The temperature dependence of the resonance γ-absorption will be determined, as in the case of a nucleus forming a part of an ideal lattice /6/, by what frequencies the impurity atom principally vibrates at. The same may also be said of the temperature shift of the Mössbauer line /7/, since it is determined by the mean kinetic energy of the absorbing (or emitting) nucleus /8/. The whole trouble in finding the intensity and temperature shift of the Mössbauer line is that we have to find the frequency spectrum of the vibrations of a nonideal crystal, and the amplitude of the vibrations of the impurity atom. There is a large amount of difficulty /5/ in solving this problem, even when the density of the impurity atoms is small.

It will be shown in the present paper, that the Green's temperature function method /9--10/ makes it possible to avoid finding the normal vibration frequencies of the nonideal crystal (with the exception of local frequencies, if there are any, and this only when finding the temperature dependence of the intensity of the Mössbauer line), and at the same time get comparatively simple, accurate expressions for the Mössbauer line intensity using various crystal models.

1. GENERAL EXPRESSION FOR THE MÖSSBAUER LINE INTENSITY.

We shall now find a general expression for the Mössbauer line intensity, assuming that the concentration of impurity atoms is small.

Let an impurity atom replace an atom of the ideal monotonic cubic lattice at the site $\underline{l} = 0$. Then, assuming that the interaction between the atoms in the ideal lattice is unchanged by replacing the atom at the site $\underline{l} = 0$ by the impurity atom of mass $\underline{m}$, we write the Hamiltonian, giving the state of the lattice, in the form

$$H = H_0 + V, \qquad (1.1)$$

where

$$H_0 = \sum_l \frac{p_l^2}{2M} + \frac{1}{2}\sum_{l,l'} u_l^\alpha A^{\alpha\alpha'}(l,l') u_{l'}^{\alpha'} ,$$

$$V = \frac{p_0^2}{2M}\sigma + \sum_l u_0^\alpha \Delta A^{\alpha\alpha'}(0,l) u_l^{\alpha'} - \frac{1}{2} u_0^\alpha \Delta A^{\alpha\alpha'}(0,0) u_0^{\alpha'} .$$

Here, H_0 is the Hamiltonian of the ideal lattice, $u_{\underline{l}}$ and $p_{\underline{l}}$ are the displacement and momentum operators of the $\underline{l}$th nucleus, $A^{\alpha\alpha'}(\underline{l},\underline{l}')$ are the forced constants of the ideal lattice, $\Delta A^{\alpha\alpha'}(0,\underline{l})$ is the change in the force constants when an atom in the ideal lattice is replaced by an impurity nucleus, $\sigma = M/m - 1$, M is the mass of an atom in the ideal lattice, and V is the perturbation produced by replacing an atom in the ideal lattice by an impurity atom. We shall assume from now on that $\Delta A^{\alpha\alpha'}(0,\underline{l}) = \varkappa A^{\alpha\alpha'}(0,\underline{l})$, where $\varkappa$ is a constant.

The Hamiltonian (1.1) is easily diagonalized /11/. Then

$$H = \sum_\gamma \varepsilon_\gamma \zeta_\gamma^+ \zeta_\gamma + \varepsilon_0 , \tag{1.2}$$

with

$$u_l^\alpha = \sum \{ W_l^\alpha(\gamma)\zeta_\gamma^+ + W_l^{\alpha *}(\gamma)\zeta_\gamma \} ,$$

where ε_γ is the energy of the elementary excitation of number γ, ε_0 is the crystal lattice energy in the ground state, ζ_γ^+ and ζ_γ are production and annihilation operators of the elementary excitations, and $W_{\underline{l}}^\alpha(\gamma)$ are some coefficients, the values of which are not important to us here.

We now take into consideration that the total effective cross-section, $\sigma(E)$, for the absorption of γ- quanta with energy E by the nucleus located at the site $\underline{l} = 0$ is represented in the form

/12--13/:

$$\sigma(E)=\frac{\Gamma\sigma_0}{4}\int_{-\infty}^{\infty}dt\,\exp\left\{it(E_0-E)-\frac{\Gamma(t)}{2}\right\}\langle e^{-iku_0(0)}e^{iku_0(t)}\rangle, \qquad (1.3)$$

where $u_\gamma^\alpha(t)=\exp(itH)u_0^\alpha(0)\exp(-itH)$ is the nuclear displacement operator in the Heisenberg representation, E_0 is the transition energy in the nucleus, Γ is the width of the excited nuclear level, σ_0 is the maximum value of the cross-section for γ-absorption by the free nucleus, $\underline{k}$ is the wave vector of the incident γ-quantum, and

$$\langle\cdots\cdots\rangle=\frac{Sp(e^{-\beta H}\cdots\cdots)}{Sp(e^{-\beta H})}, \quad \beta=\frac{1}{kT}.$$

The time correlation of the function $\langle\exp(-iku_0(0))\exp(iku_0(t))\rangle$ may be represented in the form

$$\langle e^{-iku_0(0)}e^{iku(t)}\rangle=\langle T_c\exp\left\{i\int_c dz\,\mathcal{J}^\alpha(z)u_0^\alpha(z)\right\}\rangle, \qquad (1.4)$$

where $\mathcal{J}^\alpha(z)=k^\alpha[\delta(z-t)+\delta(z)]$, T_c is the time ordering operator along the contour $\underline{c}$, shown in Fig. 1, where the point $\underline{t}$, and all the points in the upper part of the contour have the meaning of a "lower" time than the points in the lower part of the contour, in particular than the point 0.

The expression $\langle T_c\exp\{i\int_c dz\,\mathcal{J}^\alpha(z)u_0^\alpha(z)\}\rangle$ is easily evaluated for the system described by the Hamiltonian (1.1), if we use Wick's algebraic theorem /14/, generalized to the case of temperatures not equal to zero /9--15/. Actually,

$$\langle T_c\exp\left\{i\int_c\mathcal{J}^\alpha(z)u_0^\alpha(z)dz\right\}\rangle=\sum_m\frac{i^m}{m!}\int_c dz_1\cdots\int_c dz_m\,\mathcal{J}^{\alpha_1}(z_1)\cdots\mathcal{J}^{\alpha_m}(z_m)\langle T_c u_0^{\alpha_1}(z_1)\cdots u_0^{\alpha_m}(z_m)\rangle. \qquad (1.5)$$

The odd terms in this series are, from (1.2), identically equal to zero, while the even terms of the series may be transformed in such a way that the 2n term is written in the form

$$\frac{i^{2n}}{(2n)!}\sum_{\alpha_1\ldots\alpha_{2n}}\int_c dz_1\ldots\int_c dz_{2n}\, \mathcal{J}^{\alpha_1}(z_1)\ldots\mathcal{J}^{\alpha_{2n}}(z_{2n})\langle T_c u_0^{\alpha_1}(z_1)\ldots u_0^{\alpha_{2n}}(z_{2n})\rangle = \frac{1}{n}\left\{\frac{i^2}{2}\sum_{\alpha\alpha'}\int_c dz\int_c dz'\, \mathcal{J}^{\alpha}(z)\langle T_c u_0^{\alpha}(z)u_0^{\alpha'}(z')\rangle\right\}^n . \quad (1.6)$$

This is due to the fact that the number of all possible pairings of the operators $u(z_i)$ and $u(z_j)$ in the expression $\langle T_c u_0^{\alpha_1}(z_1)\ldots u^{\alpha_{2n}}(z)\rangle$ is $(2n-1)(2n-3)\ldots1$. However the use of Wick's theorem, generalized to the case $T \neq 0$, is justified by the fact that the operator $u_0^{\alpha}(z)$ is a linear function of the elementary excitation production and annihilation operators.

Using now (1.6), we represent Eq. (1.5) in the form

$$\langle T_c \exp\left\{i\int_c \mathcal{J}^{\alpha}(z)u_0^{\alpha}(z)dz\right\}\rangle = \exp\left\{-\frac{1}{2}\int_c dz\int_{c'} dz'\, \mathcal{J}^{\alpha}(z)\mathcal{J}^{\alpha'}(z')\langle T_c u_0^{\alpha}(z)u_0^{\alpha'}(z')\rangle\right\}. \quad (1.7)$$

Performing the integration over dz and dz', we obtain

$$\langle e^{iku_0(0)}e^{iku_0(t)}\rangle = \exp\left\{-k^{\alpha}k^{\alpha'}\langle u_0^{\alpha}(0)u_0^{\alpha'}(0)\rangle + k^{\alpha}k^{\alpha'}\langle u_0^{\alpha}(0)u_0^{\alpha'}(t)\rangle\right\}. \quad (1.8)$$

Thus, for the total effective γ- absorption cross-section of the impurity nucleus we obtain the expression

$$\sigma(E) = \frac{\Gamma}{4}\sigma_0\int_{-\infty}^{+\infty} dt\, e^{it(E_0-E)-\frac{\Gamma(t)}{2}}\exp\left\{-k^{\alpha}k^{\alpha'}\langle u_0^{\alpha}(0)u_0^{\alpha'}(0)\rangle + k^{\alpha}k^{\alpha'}\langle u_0^{\alpha}(0)u_0^{\alpha'}(t)\rangle\right\}. \quad (1.9)$$

The correlation function $k^{\alpha}k^{\alpha'}\langle u_0^{\alpha}(0)u_0^{\alpha'}(t)\rangle$ determines the form of the ionic part of the absorption spectrum, which, using (1.2), may be represented in the form

$$k^{\alpha}k^{\alpha'}\langle u_0^{\alpha}(0)u_0^{\alpha'}(t)\rangle = \sum_r |a_r|^2\left(\bar{n}_r e^{-i\varepsilon_r t} + (\bar{n}_r+1)e^{i\varepsilon_r t}\right), \quad (1.10)$$

where

$$\bar{n}_r = (e^{\beta \varepsilon r} - 1)^{-1} \ , \quad a_r = k^{\alpha} W^{\alpha}(r)$$

It may occur [1], that for a definite value of $\gamma = \gamma_0$ (local vibration) $|a_{\gamma_0}|^2$ is independent of N, where N is the number of elementary cells in the crystal. Then it is easily shown that the total recoilless γ absorption cross-section is equal to

$$\sigma_m(E) = \sigma_0 \frac{\exp\{-k^{\alpha}k^{\alpha'}\langle u_0^{\alpha}(0)u_0^{\alpha'}(0)\rangle\}}{(E_0-E)^2 + \frac{\Gamma}{4}} \frac{\Gamma}{4} \sum_{n=0}^{\infty} \frac{\{|a_{r_0}|^2 \sqrt{\bar{n}_{r_0}(\bar{n}_{r_0}+1)}\}^m}{m!} = \tag{1.11}$$

$$= \frac{\sigma_0 \exp\{-k^{\alpha}k^{\alpha'}\langle u_0^{\alpha}(0)u_0^{\alpha'}(0)\rangle\}}{(E_0-E)^2 + \frac{\Gamma^2}{4}} \frac{\Gamma^2}{4} I_0\left(2|a_{r_0}|^2 \sqrt{\bar{n}_{r_0}(\bar{n}_{r_0}+1)}\right) ,$$

where $I_0(z)$ is the zero order Bessel function of imaginary argument. Hence we obtain the expression for the Mössbauer factor

$$f = \exp\{-k^{\alpha}k^{\alpha'}\langle u_0^{\alpha}(0)u_0^{\alpha'}(0)\rangle\} I_0\left(2|a_r|^2 \sqrt{\bar{n}_{r_0}(\bar{n}_{r_0}+1)}\right) . \tag{1.12}$$

At zero temperature, the second factor is equal to 1, so that to find the Mössbauer line intensity at $T=0$ it is sufficient to know $\langle u_0^{\alpha}(0)u_0^{\alpha'}(0)\rangle$.

2. EQUATIONS FOR THE GREEN'S TEMPERATURE FUNCTIONS.

To evaluate $\langle u_0^{\alpha}(0)u_0^{\alpha'}(0)\rangle$ we make use of Green's temperature functions /9--10/. It is easily seen that $\langle u_0^{\alpha}(0)u_0^{\alpha'}(0)\rangle$ is a boundary value of the Green's functions, defined by the relation

$$G_{00}^{\alpha\alpha'}(\tau,\tau') = \frac{Sp\{e^{-\beta H} T \tilde{u}_0^{\alpha}(\tau)\tilde{u}_0^{\alpha'}(\tau)\}}{Sp\{e^{-\beta H}\}} = \langle T \tilde{u}^{\alpha}(\tau)\tilde{u}^{\alpha}(\tau')\rangle \cdot \tag{2.1}$$

---

1) We are grateful to M. A. Krivoglaz, for having called our attention to this fact.

where $\tilde{u}_0^{\alpha}(\tau) = \exp(\tau H)\, u_0^{\alpha}(0) \exp(-\tau H)$, where, as is easily shown, $G_{00}^{\alpha\alpha'}(\tau,\tau') = G_{00}^{\alpha\alpha'}(\tau-\tau')$, and, for $\tau-\tau' < 0$,

$$G_{00}^{\alpha\alpha'}(\tau-\tau') = G_{00}^{\alpha\alpha'}(\tau-\tau'-\beta) \tag{2.2}$$

Continuing the function (2.1) periodically over the whole straight line, we find that it holds for any values of $\tau-\tau'$.

The function (2.1) may also be written in the form

$$G_{00}^{\alpha\alpha'}(\tau,\tau') = \frac{\langle T u_0^{\alpha}(\tau) u_0^{\alpha'}(\tau') S(\beta) \rangle}{\langle S(\beta) \rangle}, \tag{2.3}$$

where

$$S(\beta) = T \exp\left\{-\int_0^{\beta} dz\, V(z)\right\}, \quad u^{\alpha}(\tau) = \exp(\tau H_0)\, u^{\alpha}(0) \exp(-\tau H_0),$$
$$V(z) = \exp(z H_0)\, V \exp(-z H_0),$$

where the statistical averaging here, and in what follows, is carried out with the Hamiltonian H_0.

In finding the integral equations which are satisfied by the Green's temperature functions, it is convenient to use the temperature form of the "generalized" Wick theorem (see /14/ p. 285), from which the equation follows for any operators A and B

$$\frac{\langle TABS(\beta)\rangle}{\langle S(\beta)\rangle} = \frac{\langle T\overline{A}B\overline{S}(\beta)\rangle}{\langle S(\beta)\rangle} + \frac{\langle T\overline{A}B\overline{S}(\beta)\rangle}{\langle S(\beta)\rangle} = \overline{AB} - \int_0^{\beta} dz \frac{\langle T\overline{A}B\overline{V}(z)S(\beta)\rangle}{\langle S(\beta)\rangle}, \tag{2.4}$$

where $\overline{AB} = \langle TAB \rangle$. Using (2.4), we obtain the following system of integral equations, which determine the function $G_{00}^{\alpha\alpha'}(\tau,\tau')$

$$G_{00}^{\alpha\alpha'}(\tau,\tau') = G_0^{\alpha\alpha'}(\tau,\tau') - \frac{\sigma}{M}\int_0^{\beta} dz\, D^{\alpha\gamma}(\tau,z) D_0^{\alpha'\gamma}(\tau',z) -$$
$$-\varkappa \int_0^{\beta} dz \left\{ X^{\alpha\gamma}(\tau,z) G_0^{\alpha'\gamma}(\tau',z) + G^{\alpha\gamma}(\tau,z)\left[X_0^{\alpha'\gamma}(\tau',z) - A(0,0) G^{\alpha'\gamma}(\tau',z)\right]\right\}. \tag{2.5}$$

$$D^{\alpha\alpha'}(\tau,\tau') = D_0^{\alpha\alpha'}(\tau,\tau') - \frac{\sigma}{M}\int_0^{\tau'} dz\, D^{\alpha\gamma}(\tau,z) F_0^{\alpha'\gamma}(\tau';z) -$$
$$- \varkappa \int_0^\beta dz \{ X^{\alpha\gamma}(\tau,z) D_0^{\gamma\alpha'}(z,\tau') + G^{\alpha\gamma}(\tau,z)[Y_0^{\gamma\alpha'}(z,\tau') - A(0,0) D_0^{\gamma\alpha'}(z,\tau')] \}, \tag{2.6}$$

$$X^{\alpha\lambda}(\tau,\tau') = X^{\alpha\lambda}(\tau,\tau') - \frac{\sigma}{M}\int_0^\beta dz\, D^{\alpha\gamma}(\tau,z) Y_0^{\lambda\gamma}(\tau';z) -$$
$$- \varkappa \int_0^\beta dz \{ X^{\alpha\gamma}(\tau,z) X_0^{\gamma\lambda}(z,\tau') + G^{\alpha\gamma}(\tau,z)[Z_0^{\gamma\lambda}(\tau';z) - A(0,0) X_0^{\gamma\lambda}(z,\tau')] \}. \tag{2.7}$$

Here we use the notation:

$$\overline{u_0^\alpha(\tau) u_0^{\alpha'}(\tau')} = G_0^{\alpha\alpha'}(\tau,\tau'), \quad \overline{u_0^\alpha(\tau) p_0^{\alpha'}(\tau')} = D_0^{\alpha\alpha'}(\tau,\tau'), \quad \overline{p_0^\alpha(\tau) p^{\alpha'}(\tau')} = F_0(\tau,\tau'),$$
$$X_0^{\alpha\alpha'}(\tau,\tau') = \sum_l A^{\alpha'\gamma}(0,l)\overline{u_0^\alpha(\tau) u_l^\gamma(\tau')}, \quad D_0^{\alpha\alpha'}(\tau,\tau') = \frac{\langle T u_0^\alpha(\tau) p_0^{\alpha'}(\tau') S(\beta)\rangle}{\langle S(\beta)\rangle} \tag{2.8}$$
$$X^{\alpha\alpha'}(\tau,\tau') = \sum_l A^{\alpha'\gamma}(0,l)\frac{\langle T u_0^\alpha(\tau) u_l^\gamma(\tau') S(\beta)\rangle}{\langle S(\beta)\rangle}, \quad Y_0^{\alpha\alpha'}(\tau,\tau') = \sum_l A^{\alpha\gamma}(0,l)\overline{u_l^\gamma(\tau) p_0^{\alpha'}(\tau')},$$
$$Z_0^{\gamma\lambda}(\tau,\tau') = \sum_l A^{\lambda\alpha'}(0,l) X_{0l}^{\alpha'\gamma}(\tau,\tau'). \quad X_{0l}^{\alpha'\gamma}(\tau,\tau') = \sum_{l'} A^{\gamma\gamma'}(0,l')\overline{u_l^{\alpha'}(\tau) u_{l'}^{\gamma'}(\tau')}$$

Now, taking Eq. (2.2) into consideration, which, as is easily shown, holds for all the functions in (2.5--2.7), we can, in these equations, make the substitution

$$\int_0^\beta dz \longrightarrow \frac{1}{2}\int_{-\beta}^\beta dz$$

After this, all the Green's functions (2.5--2.7) may be expanded in Fourier's series, for example

$$G^{\alpha\alpha'}(\tau,\tau') = \frac{1}{\beta}\sum_{n=-\infty}^\infty e^{i\varepsilon_n(\tau-\tau')} G^{\alpha\alpha'}(\varepsilon_n),$$

where

$$G^{\alpha\alpha'}(\varepsilon_n) = \frac{1}{2}\int_{-\beta}^{+\beta} e^{i\varepsilon_n\tau} G^{\alpha\alpha'}(\tau)\,d\tau \;, \quad \varepsilon_n = \frac{2\pi n}{\beta} \;, \quad n = 0, \pm 1, \pm 2 \ldots\ldots ,$$

Here, for the Fourier transforms, we obtain the equation

$$G^{\alpha\alpha'}(\varepsilon_n) = G_0^{\alpha\alpha'}(\varepsilon_n) - \frac{\sigma}{M} D^{\alpha r}(\varepsilon_n) D_0^{\alpha' r}(-\varepsilon_n) - \varkappa X^{\alpha r}(\varepsilon_n) G_0^{\alpha' r}(-\varepsilon_n) - \\ - \varkappa G^{\alpha r}(\varepsilon_n)[X_0^{\alpha' r}(-\varepsilon_n) - A(0,0) G^{\alpha' r}(-\varepsilon_n)] , \tag{2.9}$$

$$D^{\alpha\alpha'}(\varepsilon_n) = D^{\alpha\alpha'}(\varepsilon_n) - \frac{\sigma}{M} D^{\alpha r}(\varepsilon_n) F_0^{\alpha' r}(-\varepsilon_n) - \varkappa X^{\alpha r}(\varepsilon_n) D_0^{r\alpha'}(\varepsilon_n) - \\ - \varkappa G^{\alpha r}(\varepsilon_n)[Y_0^{r\alpha'}(\varepsilon_n) - A(0,0) D^{r\alpha'}(\varepsilon_n)] , \tag{2.10}$$

$$X^{\alpha\lambda}(\varepsilon_n) = X_0^{\alpha\lambda}(\varepsilon_n) - \frac{\sigma}{M} D^{\alpha r}(\varepsilon_n) Y_0^{\lambda r}(-\varepsilon_n) - \varkappa X^{\alpha r}(\varepsilon_n) X_0^{r\lambda}(\varepsilon_n) - \\ - \varkappa G^{\alpha r}(\varepsilon_n)[Z_0^{r\lambda}(-\varepsilon_n) - A(0,0) X^{r\lambda}(\varepsilon_n)] \tag{2.11}$$

Note that in deriving Eqs. (2.5--2.7) and (2.9--2.11) we have made use of the fact that in a coordinate system associated with the axes of symmetry of the crystal, $A^{\alpha\alpha'}(0,0) = \delta_{\alpha\alpha'}\, A(0,0)$.

3. MÖSSBAUER LINE INTENSITY FOR A CRYSTAL MODEL WITH DEGENERATE BRANCHES OF THE NORMAL LATTICE VIBRATIONS.

Consider the case $\varkappa = 0$. The following two equations are sufficient to find the Mössbauer line intensity in this case

$$G^{\alpha\alpha'}(\varepsilon_n) = G_0^{\alpha\alpha'}(\varepsilon_n) - \frac{\sigma}{M} D^{\alpha r}(\varepsilon_n) D_0^{\alpha' r}(-\varepsilon_n)$$
$$D^{\alpha\alpha'}(\varepsilon_n) = D_0^{\alpha\alpha'}(\varepsilon_n) - \frac{\sigma}{M} D^{\alpha r}(\varepsilon_n) F_0^{\alpha' r}(-\varepsilon_n) \,. \tag{3.1}$$

The functions $G_0^{\alpha\alpha'}(\varepsilon_n)$, $D_0^{\alpha\alpha'}(\varepsilon_n)$ and $F_0^{\alpha\alpha'}(-\varepsilon_n)$ are easily evaluated by means of (2.8) if it taken into consideration that the operators p_0^{α} and u_0^{α} are expressed in terms of the phonon production and annihilation operators $b^{+}_{g\tau}$ and $b_{g\tau}$ in the form

$$p_0^{\alpha} = i\sum_{q\tau}\left(\frac{M\hbar\omega_{q\tau}}{2N}\right)^{1/2} V^{\alpha}(q\tau)(b^{+}_{-q\tau} - b_{q\tau}) \tag{3.2}$$
$$u_0^{\alpha} = \sum_{q\tau}\left(\frac{\hbar}{2MN\omega_{q\tau}}\right)^{1/2} V^{\alpha}(q\tau)(b^{+}_{q\tau} + b_{q\tau}),$$

where $w_{g\tau}$ and $V^{\alpha}(g\tau)$ are found from the equations of motion for an ideal lattice

$$M\omega_{q\tau}^2 V^{\alpha}(q\tau) = \sum_{\alpha'} A^{\alpha\alpha'}(q) V^{\alpha'}(q\tau), \tag{3.3}$$

with the vectors $V(g\tau)$ normalized in such a way that

$$\sum_{\tau} V^{\alpha}(q\tau) V^{*\alpha'}(q\tau) = \delta_{\alpha\alpha'} \, , \quad \sum_{\alpha} V^{\alpha}(q\tau) V^{\alpha}(q\tau') = \delta_{\tau\tau'}, \tag{3.4}$$

where τ is the number of the normal vibration branch, and $\underline{g}$ is the phonon wave vector. Substituting (3.2) in (2.8), we find the following expressions for $G_0^{\alpha\alpha'}(\varepsilon_n)$, $D_0^{\alpha\alpha'}(\varepsilon_n)$ and $F_0^{\alpha\alpha'}(\varepsilon_n)$:

$$G_0^{\alpha\alpha'}(\varepsilon_n) = \frac{\hbar}{MN}\sum_{q\tau} V^{\alpha}(q\tau)V^{*\alpha'}(q\tau)\frac{\varepsilon_n}{\varepsilon_n^2 + (\hbar\omega_{q\tau})^2},$$

$$D_0^{\alpha\alpha'}(\varepsilon_n) = -\frac{\hbar}{N}\sum_{q\tau} V^{\alpha}(q\tau)V^{*\alpha'}(q\tau)\frac{\varepsilon_n}{\varepsilon_n^2 + (\hbar\omega_{q\tau})^2}, \quad (3.5)$$

$$F_0^{\alpha\alpha'}(\varepsilon_n) = \frac{M}{N}\sum_{q\tau} V^{\alpha}(q\tau)V^{*\alpha'}(q\tau)\frac{(\hbar\omega_{q\tau})}{\varepsilon_n^2 + (\hbar\omega_{q\tau})^2}$$

If $w_{q\tau}$ is independent of the branch number τ, all the Green's functions, in view of the condition (3.4), become diagonal in the indices α, α', for example,

$$G_0^{\alpha\alpha'}(\varepsilon_n) = \delta_{\alpha\alpha'} G_0(\varepsilon_n), \qquad G_0(\varepsilon_n) = \frac{\hbar^2}{MN}\sum_q \frac{1}{\varepsilon_n^2 + (\hbar\omega_q)^2}.$$

In this case, we find from Eqs. (3.1)

$$G^{\alpha\alpha'}(\varepsilon_n) = \delta_{\alpha\alpha'} G(\varepsilon_n),$$

where

$$G(\varepsilon_n) = G_0(\varepsilon_n) - \frac{\sigma}{M}\,\frac{D_0(\varepsilon_n)D_0(-\varepsilon_n)}{1 + \frac{\sigma}{M}F_0(-\varepsilon_n)},$$

$$D_0(\varepsilon_n) = -\frac{\hbar}{N}\sum_q \frac{\varepsilon_n}{\varepsilon_n^2 + (\hbar\omega_q)^2}, \quad F_0(-\varepsilon_n) = \frac{M}{N}\sum_q \frac{(\hbar\omega_q)^2}{\varepsilon_n^2 + (\hbar\omega_q)^2}. \quad (3.6)$$

For $T \longrightarrow 0$, the Mössbauer factor will be of the form

$$f = \exp\{-k^{\alpha}k^{\alpha'}\langle u_0^{\alpha}(0)u_0^{\alpha'}(0)\rangle\} \equiv \exp\{-\lim_{\beta\to\infty}\frac{k^2}{\beta}\sum_n G(\varepsilon_n)\} = f_0 \cdot f_1 \quad (3.7)$$

where

$$f_0 = \left\{ -\lim_{\beta\to\infty} \sum_q \frac{R_0}{\hbar\omega_q} (2\bar{n}_q + 1) \right. , \quad f_1 = \exp\left\{ \frac{\sigma}{M} \lim_{\beta\to\infty} \frac{k^2}{\beta} \sum_n \frac{D_0(-\varepsilon_n) D_0(\varepsilon_n)}{1 + \frac{\sigma}{M} F_0(\varepsilon_n)} \right. , \quad R_0 = \frac{(\hbar k)^2}{2M} .$$

Keeping in mind, that as $\beta \longrightarrow \infty$

$$\frac{1}{\beta} \sum_n \longrightarrow \frac{1}{2\pi} \int_{-\infty}^{+\infty} d\varepsilon$$

and using (3.6), in the Debye approximation, it is a simple matter to obtain the following final expression for the Mössbauer factor at T = 0

$$f = \exp\left\{ -\frac{3R}{2\varepsilon_0} \varphi(\sigma) \right\} , \qquad (3.8)$$

where

$$R = \frac{(\hbar k)^2}{2m} , \quad \varphi(\sigma) = (1-\sigma)^{-1} \left\{ 1 + \frac{12\sigma}{\pi} \int_0^{\pi/2} \frac{(tg^2 x + 1)(tg\, x - x)^2 dx}{tg^3 x \left[(1+\sigma) tg^3 x - 3\sigma (tg\, x - x) \right]} \right\} , \quad \varepsilon_0 = \hbar\omega_{max} .$$

The results of a numerical calculation[2)] of $\varphi(\sigma)$ as a function of the mass ratio M/m is shown in Fig. 2. It may be seen from the figure that to get a large Mössbauer line intensity, the impurity atoms should be put into a heavy lattice (assuming that the solvent crystals are at the same Debye temperature). The best solvent crystals will be those with atoms of large mass and with large De-

---

2) We are grateful to N. N. Matviishina, for making the numerical calculations on the computing machine of the Cybernetic Institute of the Academy of Sciences, UkrSSR.

bye temperature.

The Mössbauer factor for impurity atoms, calculated from the above dependence on the ratio of the masses of the atoms in the impurity and the solvent, are in qualitative agreement with the existing experimental data /16--17/.

We shall now give a more detailed discussion of the case $\sigma < 0$ (local frequencies absent). The expression (3.6) for $G(\varepsilon_n)$ may be written in the form

$$G(\varepsilon_n) = \frac{\hbar^2}{m} \qquad \frac{\tau(\varepsilon_n)}{1+\sigma[1-\varepsilon_n^2\tau(\varepsilon_n)]}, \qquad (3.9)$$

where

$$\tau(\varepsilon_n) = \frac{1}{N}\sum_q \frac{1}{\varepsilon_n^2+\varepsilon_q^2}, \qquad \varepsilon_q = \hbar\omega_q .$$

Accordingly, for the Mössbauer factor we obtain the expression

$$f = \exp\left\{-\frac{k^2}{\beta}\sum_n G(\varepsilon_n)\right\} \equiv \exp\left\{-\frac{2R}{\beta}\sum_n \frac{\tau(\varepsilon_n)}{1+\sigma-\sigma\varepsilon_n^2\tau(\varepsilon_n)}\right\}. \qquad (3.10)$$

In the limiting case of a very heavy impurity nucleus, where $\sigma \to -1$, and at low temperatures, the principal contribution to the sum over n is made by the small ε_n's. Accordingly, in this case

$$\frac{1}{\beta}\sum_n \frac{\tau(\varepsilon_n)}{1+\sigma-\sigma\varepsilon_n^2\tau(\varepsilon_n)} \simeq \frac{1}{(-\sigma\varepsilon')}\left[(e^{\beta\varepsilon'}-1)^{-1}+\frac{1}{2}\right],$$

where

$$\varepsilon' = \left(\frac{1+\sigma}{-\sigma\tau(0)}\right)^{1/2}.$$

Thus

$$f = \exp\left\{\frac{R}{\sigma\varepsilon'}\left[2\left(e^{\beta\varepsilon'}-1\right)^{-1}+1\right]\right\}, \tag{3.11}$$

in particular, at T = 0

$$f = \exp\left\{-R\left(\frac{m}{M}\right)^{1/2}(-\sigma)^{-1/2}\tau^{1/2}(0)\right\}. \tag{3.12}$$

It follows from (3.4) that even at temperatures satisfying the condition $\beta\varepsilon' < 1$ (in the Debye model, this is equivalent to the condition $T > (M/m)^{\frac{1}{2}}\Theta$), the Mössbauer line intensity drops off exponentially with increase in temperature. This conclusion is in agreement with the results obtained in /5/.

At high temperatures (in the Debye model, $T > \Theta$, the Mössbauer line intensity is given by the expression

$$f = \exp\left\{-\frac{(\hbar k)^2}{M}(1+\sigma)\frac{1}{\beta}\sum_n \frac{\tau(\varepsilon_n)}{1+\sigma-\sigma\varepsilon_n^2\,\tau(\varepsilon_n)}\right\} \simeq \exp\left\{-\frac{2R_0}{\beta}\tau(0)\right\}, \tag{3.13}$$

which is independent of the mass of the impurity nucleus.

To find the Mössbauer line intensity for the case $\sigma > 0$, when there is a local frequency in the normal vibration frequency spectrum of the ideal lattice, we must find the quantity $|a_{\gamma 0}|^2$, corresponding to the contribution made by the local frequency to the time correlation function $\langle u_0^{\alpha}(t)u_0^{\alpha'}(t')\rangle$. It is known that the spectral density $\mathcal{J}_{\alpha\alpha'}(\varepsilon)$, giving the time correlation function $\langle u_0^{\alpha}(t)u_0^{\alpha'}(t')\rangle$ with the aid of the relation

$$\langle u_0^{\alpha}(t)u_0^{\alpha'}(t')\rangle = \int_{-\infty}^{\infty} d\varepsilon\, e^{-i\varepsilon(t-t')}\mathcal{J}_{\alpha\alpha'}(\varepsilon), \tag{3.14}$$

is equal to

$$\mathcal{J}_{\alpha\alpha'}(\varepsilon) = \frac{1}{i}\left\{K_r^{\alpha\alpha'}(\varepsilon + i\delta) - K_a^{\alpha\alpha'}(\varepsilon - i\delta)\right\}, \qquad (3.15)$$

where $\mathcal{K}^{\alpha\alpha'}_{\tau, a}(\varepsilon)$ is the analytic continuation of the Fourier transforms of the functions

$$K_r^{\alpha\alpha'}(t,t') = i\theta(t-t')\langle[u^{\alpha}(t), u^{\alpha'}(t')]\rangle,$$
$$K_a^{\alpha\alpha'}(t,t') = -i\theta(t'-t)\langle[u^{\alpha}(t), u^{\alpha'}(t')]\rangle,$$

in the upper and lower half plane ε (18) respectively.

On the other hand, it is easily shown /10/ that the functions $K^{\alpha\alpha'}_{\tau}(\varepsilon)$ and $K^{\alpha\alpha'}_{a}(\varepsilon)$ are the analytic continuation of the Fourier transform of the Green's temperature function $G^{\alpha\alpha'}(\varepsilon_{\gamma})$ in the upper and lower half plane respectively, i.e., we have the equation

$$K_r^{\alpha\alpha'}(\varepsilon) = \frac{1}{2\pi} G^{\alpha\alpha'}(-i\varepsilon), \qquad \mathrm{Im}\,\varepsilon > 0$$
$$K_a^{\alpha\alpha'}(\varepsilon) = \frac{1}{2\pi} G^{\alpha\alpha'}(-i\varepsilon), \qquad \mathrm{Im}\,\varepsilon < 0$$

so that, making use of (3.9), we obtain

$$K_{\tau,a}^{\alpha\alpha'}(\varepsilon) = \delta_{\alpha\alpha'}\frac{\hbar^2}{2\pi m}\;\frac{\tau_1(\varepsilon)}{1+\sigma\tau_2(\varepsilon)} \qquad \mathrm{Im}\,\varepsilon \neq 0, \qquad (3.16)$$

$$\tau_1(\varepsilon) = \frac{1}{N}\sum_q \frac{1}{-\varepsilon^2+\varepsilon_q^2}, \qquad \tau_2(\varepsilon) = \frac{1}{N}\sum_q \frac{\varepsilon_q^2}{-\varepsilon^2+\varepsilon_q^2}.$$

The functions $K_{\gamma}^{\alpha\alpha'}(\mathcal{E})$ and $K_{a}^{\alpha\alpha'}(\mathcal{E})$ form one function, analytic in the whole plane of $\mathcal{E}$, with the exception of a possible section on the real axis /10/. The poles of this function on the real axis determine the spectrum of the elementary excitations of the system. In our case, to find the normal vibration frequencies of the nonideal crystal, we obtain the equation

$$1+\sigma\,\tau_2(\varepsilon)=0\,, \tag{3.17}$$

which is similar to the equation obtained in /4,5/. It follows from (3,17) that only for $\sigma > 0$ is there a local frequency in the normal vibration frequency spectrum of the nonideal crystal.

To find the time correlation function, we substitute (3.16) and (3.15) in (3.14). Then, evaluating the integral (3.14) by the subtraction method, we obtain

$$\langle u_0^{\alpha}(t)\,u_0^{\alpha}(t')\rangle = \langle u^{\alpha}(t)u^{\alpha}(t')\rangle' - \Big\{[\bar{n}(\varepsilon_0')+1]\,e^{-i\varepsilon_0'(t-t')} + \bar{n}(\varepsilon_0')\,e^{i\varepsilon_0'(t-t')}\Big\}\,\delta_{\alpha\alpha'}\,\frac{\hbar^2}{m}\,\frac{\tau_1(\varepsilon_0')}{2\varepsilon_0'\,\sigma\left(\frac{d\tau_2}{d\varepsilon^2}\right)_{\varepsilon=\varepsilon_0'}}\,, \tag{3.18}$$

where $\bar{n}(\mathcal{E}_0') \doteq (e^{\beta\mathcal{E}_0'}-1)^{-1}$, and $\mathcal{E}_0'$ is the energy of the local vibration, found from Eq. (3.17). The function $\langle u^{\alpha}(t)u^{\alpha'}(t')\rangle'$ gives the contribution to the time correlation function of all the frequencies, with the exception of the local frequency (in the model under discussion, triply degenerate). The expression (3.18) may also be rewritten, by using (3.17), in the form

$$\langle u_0^{\alpha}(t)u_0^{\alpha'}(t')\rangle = \delta_{\alpha\alpha'}\,\frac{\hbar^2(1+\sigma)}{2m\varepsilon_0'}\,\frac{\partial \ln(\varepsilon_0')^2}{\partial\sigma}\Big\{\bar{n}(\varepsilon_0')\,e^{i\varepsilon_0'(t-t')} + [\bar{n}(\varepsilon_0')+1]\,e^{-i\varepsilon_0'(t-t')}\Big\} + \langle u^{\alpha}(t)\,u^{\alpha'}(t')\rangle'\,. \tag{3.19}$$

By comparing (3.19) with (1.10), we obtain for $|a_{\varepsilon_0'}|^b$ the expression

$$|a_{\varepsilon_0'}|^2 = \frac{R}{\varepsilon_0'}(1+\sigma)\frac{\partial l_n(\varepsilon_0')^2}{\partial\sigma}. \qquad (3.19a)$$

For the case where $\varepsilon_0' \gg \varepsilon_0$, the value of the local frequency may be found from Eq. (3.17), writing $\tau_2(\varepsilon)$ in the form

$$\tau_2(\varepsilon) = -\overline{\varepsilon^2}/(\varepsilon_0')^2 - \cdots\cdots\cdots \qquad \text{where} \quad \overline{\varepsilon^2} = \frac{1}{N}\sum_q \varepsilon_q^2 .$$

when $(\varepsilon_0')^2 = \sigma\overline{\varepsilon^2} + \cdots\cdots\cdots$ and $|a_{\varepsilon_0'}|^2 = \frac{R}{\varepsilon_0'}\frac{1+\sigma}{\sigma}$.

For the Debye model, ε_0' may be found by solving Eq. (3.17) numerically.

Thus, the Mössbauer line intensity is given, for the case under discussion ($\sigma > 0$), by the expression

$$f = \exp\left\{-2R\frac{1}{\beta}\sum_n \frac{g(\varepsilon_n)}{1+\sigma-\sigma\varepsilon_n^2 g(\varepsilon_n)}\right\} I_0\left\{\frac{2R}{\varepsilon_0'}(1+\sigma)\sqrt{\bar{n}(\varepsilon_0')[\bar{n}(\varepsilon_0')+1]}\frac{\partial l_n(\varepsilon_0')^2}{\partial\sigma}\right\}, \qquad (3.20)$$

where ε_0' is found from Eq. (3.17).

We pass, finally, to a discussion of the general case, where we have both $\sigma \neq 0$ and $\varkappa \neq 0$. For a crystal model with three degenerate branches of the normal vibration frequency spectrum, Eqs. (2.9), (2.10), and (2.11) take the form

$$G(\varepsilon_n)[1+\varkappa\bar{X}_0(\varepsilon_n)] = G_0(\varepsilon_n) - \frac{\sigma}{M}D(\varepsilon_n)D_0(-\varepsilon_n) - \varkappa X(\varepsilon_n)G_0(\varepsilon_n), \qquad (3.21)$$

$$-G(\varepsilon_n)\varkappa\bar{Y}_0(\varepsilon_n) = -D_0(\varepsilon_n) + D(\varepsilon_n)\left[1+\frac{\sigma}{M}F_0(-\varepsilon_n)\right] + \varkappa X(\varepsilon_n)D_0(\varepsilon_n), \qquad (3.22)$$

$$-G(\varepsilon_n)\varkappa\bar{Z}_0(\varepsilon_n) = -\bar{X}_0(\varepsilon_n) + \frac{\sigma}{M}D(\varepsilon_n)Y_0(-\varepsilon_n) + [1+\varkappa X_0(\varepsilon_n)]X(\varepsilon_n), \qquad (3.23)$$

where

$$\bar{X}_0(\varepsilon_n) = X_0(\varepsilon_n) - A(0,0)G_0(\varepsilon_n), \quad \bar{Y}_0(\varepsilon_n) = Y_0(\varepsilon_n) - A(0,0)D_0(\varepsilon_n),$$
$$\bar{Z}_0(\varepsilon_n) = Z_0(\varepsilon_n) - A(0,0)X_0(\varepsilon_n), \quad G^{\alpha\alpha'}(\varepsilon_n) = \delta_{\alpha\alpha'}G(\varepsilon_n) .$$

Keeping in mind Eq. (3.3), and using the expression (3.5) we find

$$\begin{aligned}
&X_0(\varepsilon_n) = 1 - \varepsilon_n^2 \tau(\varepsilon_n); \qquad Y_0(\varepsilon_n) = -\frac{M\varepsilon_n}{\hbar}\left[1 - \varepsilon_n^2 \tau(\varepsilon_n)\right],\\
&\bar{X}_0(\varepsilon_n) = 1 - (\varepsilon_n^2 - \bar{\varepsilon}^2)\tau(\varepsilon_n), \quad \bar{Y}_0(\varepsilon_n) = -\frac{M\varepsilon_n}{\hbar}\left[1 - (\varepsilon_n^2 + \bar{\varepsilon}^2)\tau(\varepsilon_n)\right],\\
&\bar{Z}_0(\varepsilon_n) = \frac{M\varepsilon_n^2}{\hbar}\left[(\varepsilon_n^2 + \bar{\varepsilon}^2)\tau(\varepsilon_n) - 1\right], \quad G_0(\varepsilon_n) = \frac{\hbar^2}{M}\tau(\varepsilon_n),\\
&D_0(\varepsilon_n) = -\hbar\varepsilon_n\tau(\varepsilon_n), \quad F_0(\varepsilon_n) = M\left[1 - \varepsilon_n^2\tau(\varepsilon_n)\right],
\end{aligned} \tag{3.24}$$

where

$$\bar{\varepsilon}^2 = \hbar^2 A(0,0), \qquad \tau(\varepsilon_n) = \frac{1}{N}\sum_q \frac{1}{\varepsilon_n^2 + \varepsilon_q^2} .$$

To calculate the function $G(\varepsilon_n)$ we first eliminate $X(\varepsilon_n)$ from Eqs. (3.21) and (3.22). This gives the equation

$$-\varepsilon_n G(\varepsilon_n) = \frac{\hbar}{M}(1+\sigma)D(\varepsilon_n) .$$

On the other hand by eliminating $X(\varepsilon_n)$ from Eqs. (3.22) and (3.23), we find

$$\varkappa(1+\varkappa)\frac{M\varepsilon_n}{\hbar}\left[1 - (\varepsilon_n^2 + \bar{\varepsilon}^2)\right]G(\varepsilon_n) = D(\varepsilon_n)\left\{1 + \left[1 - \varepsilon_n^2\tau(\varepsilon_n)\right]\times\right.$$
$$\left.\times\left[\varkappa + \sigma(1+\varkappa)\right]\right\} - D_0(\varepsilon_n) .$$

Solving this system for $G(\varepsilon_n)$, we obtain the expression

$$G(\varepsilon_n) = \frac{\hbar^2(1+\sigma)}{M(1+\varkappa)}\tau(\varepsilon_n)\left\{(1+\sigma)(1+\varkappa) - \left[\varkappa(1+\sigma) + \sigma + \frac{\varkappa}{1+\varkappa}\right]\varepsilon_n^2\tau(\varepsilon_n) - \varkappa(1+\sigma)\bar{\varepsilon}^2\tau(\varepsilon_n)\right\}^{-1} .$$

If there is a local frequency in the normal vibration frequency spectrum, the contribution that it makes to the time correlation function is given by the coefficient $|a_{\varepsilon'_0}|^2$, which, for the case $\varkappa \neq 0$, is given by Eq. (3.19a), the only difference being that the local frequency is found from the equation

$$(1+\sigma)(1+\varkappa)+\left[\varkappa(1+\sigma)+\sigma+\frac{\varkappa}{1+\varkappa}\right](\varepsilon'_0)^2\tau_1(\varepsilon'_0)-\varkappa(1+\sigma)\overline{\varepsilon^2}\,\tau_1(\varepsilon'_0)=0 \quad . \tag{3.25}$$

Thus, the Mössbauer line intensity is given in the general case by the expression

$$f=I_0\left(\frac{2R(1+\sigma)}{\varepsilon'_0}\sqrt{\bar{n}(\varepsilon'_0)[\bar{n}(\varepsilon'_0)+1]}\;\frac{\partial \ln(\varepsilon'_0)^2}{\partial\sigma}\right)\exp\Big\{-\sum_n\Big[(1+\sigma)(1+\varkappa)- \tag{3.26}$$
$$-\left(\varkappa(1+\sigma)+\sigma+\frac{\varkappa}{1+\varkappa}\right)\varepsilon_n^2\,\tau(\varepsilon_n)-(1+\sigma)\varkappa\,\overline{\varepsilon^2}\,\tau(\varepsilon_n)\Big]^{-1}\frac{2R}{\beta(1+\varkappa)}\Big\},$$

which may be used in numerical calculations, for example, in the Debye model.

For zero temperature, the second factor in (3.26) is equal to 1, and the sum over n may be replaced by the integral. The value of the local frequency, for arbitrary values of σ and $\varkappa$, with a definite crystal model, may be calculated numerically from Eq. (3.25). In some limiting cases, Eq. (3.26) may also be evaluated, but we shall not stop to discuss this here. We shall point out only that the Green's temperature function method makes it possible to find the Mössbauer line intensity for lattices that are even more complicated than those discussed in the paper.

Institute of Physics, Academy of Sciences, UkrSSR

REFERENCES

/1/ H. B. Rosenstock, G. C. Klick, Phys. Rev. 119, 1198, 1960

/2/ R. F. Wallis, A. A. Maradudin, Progr. Theor. Phys. 24, 1055 1960

/3/ A. A. Maradudin, P. Mazur, E. W. Montroll, G. H. Weiss, Rev. Mod. Phys. 30, 175, 1958

/4/ I. M. Lifshits, Nuova Cimento 3 Suppl. 4, 716, 1956

/5/ Yu. Kagan, Ya. I. Iosilevskii, ZhÉTF, 42, 259, 1962

/6/ Yu. Kagan, ZhÉTF, 41, 659, 1961

/7/ R. V. Pound, G. A. Rebka, Phys. Rev. Lett. 4, 274, 1960

/8/ B. O. Josephson, Phys. Rev. Lett. 4, 341, 1960

/9/ J. Matsubara, Progr. Theor. Phys. 14, 351, 1955

/10/ A. A. Abrikosov, L. P. Gor'kov, I. E. Dzyaloshchinskii, ZHÉTF 36, 900, 1959

/11/ M. M. Bogolyubov, Lectures on Quantum Statistics /in Russian/ Kiev, 1949

/12/ I. P. Dzyub, A. F. Lubchenko, Izv. AN SSSR, Physics series 25, 893, 1961

/13/ K. S. Siugwi and A. Sjölander, Phys. Rev. 120, 1093, 1960

/14/ N. N. Bogolyubov and D. V. Shirkov, Introduction to the Theory of Quantum Fields /in Russian/ GITTL, M. 1957

/15/ D. J. Thouless, Phys. Rev. 107, 1162, 1957

/16/ D. A. Shirley and Kaplan, Phys. Rev. 123, 816, 1961

/17/ S. P. Schiffer, S. Heberle and P. Parks, Bull. Am. Phys. Soc. 6, 442, 1961

/18/ N. N. Bogolyubov, S. V. Tyablikov, DAN SSSR, 126, 53, 1959

Translated by: Charles V. Larrick.

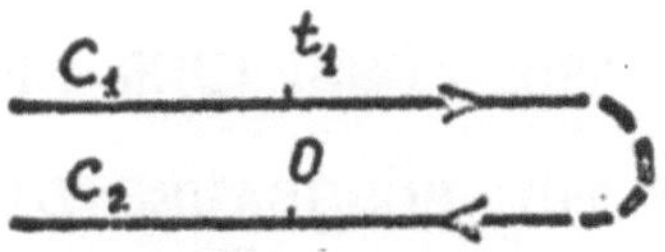

Fig. 1.

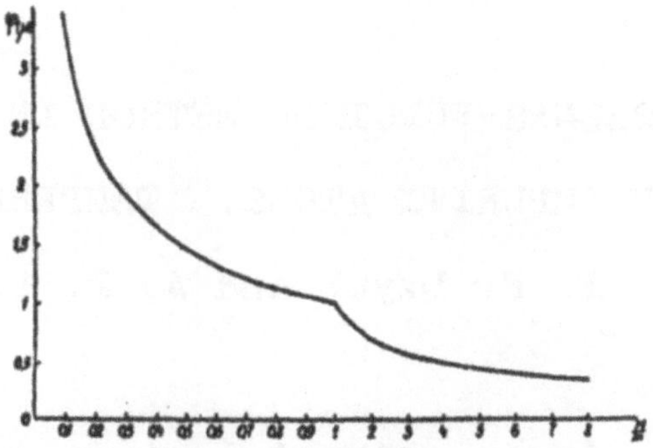

Fig. 2.

THE GREEN'S-FUNCTION METHOD IN THE THEORY OF THE MÖSSBAUER EFFECT ON IMPURITY ATOMS. TEMPERATURE SHIFT OF THE MÖSSBAUER LINE

I. P. Dzyub and A. F. Lubchenko

An experimental study of the temperature shift of the Mössbauer line for Fe^{57} nuclei present in a regular lattice, and a theoretical explanation of this effect, have been presented in /1,2/. The theoretical explanation of the effect is that as the result of emission (or absorption) of a γ-ray quantum by a nucleus the mass of the nucleus changes by the amount $\delta m = E_0/c^2$ (E_0 is the energy of the transition in the nucleus). This leads to a change of the energy of the lattice in a no-phonon transition, and this in turn is compensated by a change of the energy of the γ-ray quantum.

The temperature shift of the Mössbauer line for impurity nuclei has been studied experimentally and theoretically in /3,4/. In /4/ analytic expressions for low and high temperatures were obtained for a simple model of the lattice.

In the present paper it will be shown that by means of the method of temperature-dependent Green's functions one can in the general case determine the temperature shift of the Mössbauer line. For the model of a monatomic crystal with degenerate branches of the spectrum of normal vibrations we give the results of a numerical calculation of the temperature dependence of the shift of the Mössbauer line for various ratios of the mass of an impurity atom to that of an atom of the regular lattice.

1. STATEMENT OF THE PROBLEM

For simplicity we consider a monatomic crystal in which some of the atoms are replaced by impurity atoms. We shall suppose that the concentration of impurity atoms is small, so that in what follows we shall consider a macroscopic crystal in which there is one impurity atom of mass m at the site $\ell = 0$.

We further assume that the interaction between the atoms of the regular lattice is not changed by the replacement of the atom at the site $\ell = 0$ by an impurity atom, and set up the Hamiltonian describing the crystal lattice with one impurity atom in the form

$$H = H_0 + V$$

$$H_0 = \sum_\ell \frac{p_\ell^2}{2M} + \frac{1}{2}\sum u_\ell^\alpha A^{\alpha\alpha'}(\ell,\ell') u_{\ell'}^{\alpha'} \tag{1.1}$$

$$V = \frac{p_0^2}{2M}\mu + \sum_\ell u_0^\alpha \Delta A^{\alpha\alpha'}(0,\ell) u_\ell^{\alpha'} - \frac{1}{2} u_0^\alpha \Delta A^{\alpha\alpha'}(0,0) u_0^{\alpha'},$$

where u_ℓ^α and p_ℓ^α are respectively the displacement of the nucleus from its equilibrium position and its momentum, M is the mass of an atom of the regular lattice, $A^{\alpha\alpha'}(\ell, \ell')$ is the dynamical interaction matrix of the regular lattice, $\Delta A^{\alpha\alpha'}(0, \ell)$ is the change of the dynamical matrix owing to the replacement of an atom of the regular lattice by an impurity atom, and $\mu = \frac{M}{m} - 1$.

In what follows we assume that $\Delta A^{\alpha\alpha'}(0,\ell) = \varkappa A^{\alpha\alpha'}(0,\ell)$, where $\varkappa$ is a constant. In the expression (1.1) H_0 describes the vibrations of the atoms of the regular lattice, and V is the perturbation in the lattice caused by replacement of an atom of the regular lattice by an impurity atom.

The temperature shift of the Mössbauer line is given by /2/

$$\delta E = \langle \delta H \rangle = \langle \delta \frac{p_0^2}{2m} \rangle = -\frac{\delta m}{m} \langle \frac{p_0^2}{2m} \rangle = -\frac{E_0}{mc^2} \langle \frac{p_0^2}{2m} \rangle, \tag{1.2}$$

where E_0 is the energy of the transition in the nucleus, and

$$\langle \cdots \rangle = \frac{Sp(e^{-\beta H}\cdots)}{Sp(e^{-\beta H})}, \quad \beta = \frac{1}{K_B T}.$$

Thus to determine the temperature shift of the Mössbauer line we must know the mean value

$$\langle p_0^{\alpha}(0)\, p_0^{\alpha'}(0) \rangle \qquad , \alpha, \alpha' = 1, 2, 3.$$

It is easy to see that this average is the limiting value of the following temperature-dependent Green's function /5/

$$F^{\alpha\alpha'}(\tau,\tau') = \langle T \tilde{p}_0^{\alpha}(\tau) \tilde{p}_0^{\alpha'}(\tau') \rangle \tag{1.3}$$

$$\tilde{p}_0^{\alpha}(\tau) = exp(\tau H)\, p_0^{\alpha}(0)\, exp(-\tau H).$$

We note the following important properties of the function $F^{\alpha\alpha'}(\tau,\tau')$

$$F^{\alpha\alpha'}(\tau,\tau') = F^{\alpha\alpha'}(\tau-\tau') \tag{1.4}$$

$$F^{\alpha\alpha'}(\tau-\tau') = F^{\alpha\alpha'}(\tau-\tau'+\beta) \qquad , \tau-\tau' < 0. \tag{1.5}$$

We shall take the function $F^{\alpha\alpha'}(\tau - \tau')$ as defined in the interval $(-\beta, \beta)$ by the expression (1.3). In what follows it is convenient to continue it periodically into the entire line. Then the relation (1.5) will hold for arbitrary $\tau - \tau'$ /6/. In the interaction representation the function $F^{\alpha\alpha'}(\tau, \tau')$ has the form

$$F^{\alpha\alpha'}(\tau,\tau') = \frac{\langle T p_0^{\alpha}(\tau) p_0^{\alpha'}(\tau') S(\beta) \rangle}{\langle S(\beta) \rangle} \tag{1.6}$$

$$S(\beta) = T exp\left(-\int_0^{\beta} dz\, V(z)\right)$$

$$V(z) = exp(z H_0)\, V exp(-z H_0)$$

$$p_0^{\alpha}(\tau) = exp(\tau H_0)\, p_0^{\alpha}(0)\, exp(-\tau H_0).$$

Here and in what follows the statistical averages are taken with the Hamiltonian H_0. We define the function $F^{\alpha\alpha'}(\tau, \tau')$ by the equation that it satisfies.

2. EQUATIONS FOR THE TEMPERATURE-DEPENDENT GREEN'S FUNCTIONS AND THE TEMPERATURE SHIFT OF THE MÖSSBAUER LINE

To obtain the integral equations which are satisfied by the temperature-dependent Green's functions, we use a temperature modification of the "generalized" Wick theorem (cf. /7/, p. 285), from which we have the following equation for any operators A and B.

$$\frac{\langle TABS(\beta)\rangle}{\langle S(\beta)\rangle} = \frac{\langle T\overline{AB}\, S(\beta)\rangle}{\langle S(\beta)\rangle} + \frac{\langle TA\overline{BS}(\beta)\rangle}{\langle S(\beta)\rangle} = \overline{AB} - \int_0^\beta dz \frac{\langle TA\overline{BV}(z)S(\beta)\rangle}{\langle S(\beta)\rangle}, \qquad (2.1)$$

where $\overline{AB} = \langle TAB\rangle$.

Let us separately consider two cases:

1) $\mu \neq 0, \quad \varkappa = 0$

2) $\mu \neq 0, \quad \varkappa \neq 0$

In the case $\varkappa = 0$ we can easily write the equation for $F^{\alpha\alpha'}(\tau, \tau')$ by using the relation (2.1). In fact, we have

$$\frac{\langle Tp_0^{\alpha}(\tau)p_0^{\alpha'}(\tau')S(\beta)\rangle}{\langle S(\beta)\rangle} = \overline{p_0^{\alpha}(\tau)p_0^{\alpha'}(\tau')} - \frac{\mu}{M}\int_0^\beta dz \frac{\langle Tp_0^{\alpha}(\tau)p_0^{\gamma}(z)S(\beta)\rangle}{\langle S(\beta)\rangle}\overline{p_0^{\gamma}(z)p_0^{\alpha'}(\tau')} \qquad (2.2)$$

or, introducing the notation

$$F_0^{\alpha\alpha'}(\tau,\tau') = \overline{p_0^{\alpha}(\tau)p_0^{\alpha'}(\tau')} \equiv \langle Tp_0^{\alpha}(\tau)p_0^{\alpha'}(\tau')\rangle, \qquad (2.2a)$$

we can rewrite the equation (2.2) in the form

$$F^{\alpha\alpha'}(\tau,\tau') = F_0^{\alpha\alpha'}(\tau,\tau') - \frac{\mu}{M}\int_0^\beta dz\, F^{\alpha\gamma}(\tau,z)\, F_0^{\gamma\alpha'}(z,\tau'). \qquad (2.3)$$

Using the relation (1.5), we can make the replacement

$$\int_0^\beta dz \ldots\ldots \longrightarrow \frac{1}{2}\int_{-\beta}^\beta dz \ldots .$$

in Eq. (2.3). When we now expand the functions $F^{\alpha\alpha'}(\tau, \tau')$, $F_0^{\alpha\alpha'}(\tau, \tau')$ in Fourier series according to the formulas

$$F^{\alpha\alpha'}(\tau,\tau') = \frac{1}{\beta}\sum_{n=-\infty} e^{-i\varepsilon_n(\tau-\tau')} F^{\alpha\alpha'}(\varepsilon_n)$$

$$F^{\alpha\alpha'}(\varepsilon_n) = \frac{1}{2}\int_{-\beta}^{\beta} d\tau\, e^{i\varepsilon_n\tau} F^{\alpha\alpha'}(\tau) \quad , \varepsilon_n = \frac{2\pi n}{\beta},\ n = 0, \pm 1, \pm 2, \ldots\ ,$$

we get an equation for the Fourier transform of the function $F^{\alpha\alpha'}(\tau, \tau')$:

$$F^{\alpha\alpha'}(\varepsilon_n) = F_0^{\alpha\alpha'}(\varepsilon_n) - \frac{\mu}{M} \mp^{\alpha\gamma}(\varepsilon_n) F_0^{\gamma\alpha'}(\varepsilon_n) \ . \qquad (2.4)$$

The equation (2.4) can be solved for any type of crystal if the function $F^{\alpha\alpha'}(\epsilon_n)$, which describes the properties of the regular crystal lattice, is known.

To determine the Green's function $F_0^{\alpha\alpha'}(\epsilon_n)$ we shall start from the fact that for the regular lattice the momentum operator p_0^α can be represented in terms of the creation operators $b^+_{q\sigma}$ and destruction operators $b_{q\sigma}$ of the phonons (q is the momentum of the phonon and σ is the number of the branch of normal vibrations of the regular lattice) in the following way:

$$p_0^\alpha = \frac{i}{N^{1/2}} \sum_{q\sigma} \left(\frac{M\hbar\omega_{q\sigma}}{2}\right)^{1/2} V^\alpha(q\sigma)(b^+_{q\sigma} - b_{q\sigma}) \ , \qquad (2.5)$$

where N is the number of elementary cells in the crystal and $V^\alpha(q\sigma)$ satisfies the equation

$$M\omega^2_{q\sigma} V^\alpha(q\sigma) = A^{\alpha\alpha'}(q) V^{\alpha'}(q\sigma),\ A^{\alpha\alpha'}(q) = \sum_{l-l'} e^{-iq(l-l')} A^{\alpha\alpha'}(l,l')$$

and the orthonormality conditions

$$\sum_\sigma V^\alpha(q\sigma) V^{*\alpha'}(q\sigma) = \delta_{\alpha\alpha'}$$
$$\sum_\alpha V^\alpha(q\sigma) V^{*\alpha}(q\sigma') = \delta_{\sigma\sigma'} \ . \qquad (2.6)$$

Keeping in mind the definition (2.2a) and Eq. (2.5), we easily find for the function $F_0^{\alpha\alpha'}(\tau, \tau')$ the expression $F_0^{\alpha\alpha'}(\tau,\tau')$

$$F_0^{\alpha\alpha'}(\tau,\tau') = \frac{1}{N}\sum_{q\sigma} \frac{M\hbar\omega_{q\sigma}}{2} V^\alpha(q\sigma) V^{*\alpha'}(q\sigma)\left[\bar{n}_{q\sigma} e^{\hbar\omega_{q\sigma}|\tau-\tau'|} + (\bar{n}_{q\sigma}+1) e^{-\hbar\omega_{q\sigma}|\tau-\tau'|}\right],$$

$$\bar{n}_{q\sigma} = (e^{\beta\hbar\omega_{q\sigma}} - 1)^{-1}$$

and for its Fourier transform the expression

$$F_0^{\alpha\alpha'}(\varepsilon_n) = \frac{M}{N}\sum_{q\sigma} V^\alpha(q\sigma) V^{*\alpha'}(q\sigma) \frac{(\hbar\omega_{q\sigma})^2}{\varepsilon_n^2 + (\hbar\omega_{q\sigma})^2} \ .$$

Ordinarily the vibration vectors $V^{\alpha}(q\sigma)$ of the regular lattice are unknown. To estimate the effects expected here let us consider the model of a crystal with degenerate branches of normal lattice vibrations, i. e., we suppose that $\omega_{q\sigma} = \omega_q$; then, using Eq. (2.6), we get

$$F_0^{\alpha\alpha'}(\varepsilon_n) = \delta_{\alpha\alpha'} \frac{M}{N} \sum_q \frac{(\hbar\omega_q)^2}{\varepsilon_n^2 + (\hbar\omega_q)^2} = \delta_{\alpha\alpha'} \cdot F_0(\varepsilon_n). \tag{2.7}$$

Solving the equation (2.4), we have as the expression for $F^{\alpha\alpha'}(\epsilon_n)$

$$F^{\alpha\alpha'}(\varepsilon_n) = \delta_{\alpha\alpha'} \frac{F_0(\varepsilon_n)}{1 + \frac{\mu}{M} F_0(\varepsilon_n)} . \tag{2.8}$$

From this we can easily obtain the temperature shift of the Mössbauer line. Using Eqs. (1.2), (1.3), and (2.8), we get

$$\delta E = -\frac{3E_0 M}{2m^2c^2} \frac{1}{\beta} \sum_n \frac{f(\varepsilon_n)}{1+\mu f(\varepsilon_n)} , \quad f(\varepsilon_n) = \frac{1}{N} \sum_q \frac{\varepsilon_q^2}{\varepsilon_n^2 + \varepsilon_q^2} , \quad \varepsilon_q = \hbar\omega_q . \tag{2.9}$$

In the Debye approximation, with the frequency distribution of the normal vibrations given by the function

$$g(\varepsilon_q) = \frac{3}{\varepsilon_0^3} \varepsilon_q^2 , \qquad \varepsilon_0 = k_\beta \Theta ,$$

where θ is the Debye temperature, we have as the expression for $f(\epsilon_n)$:

$$f(\varepsilon_n) = 1 - 3\left(\frac{\varepsilon}{\varepsilon_0}\right)^2 \left[1 - \frac{\varepsilon_n}{\varepsilon_0} \operatorname{arctg} \frac{\varepsilon_0}{\varepsilon_n}\right] .$$

In this case we can put the expression for the temperature shift in the form

$$\delta E = -\frac{2E_0}{2mc^2} \varepsilon_0 \chi(\xi) , \quad \chi(\xi) = \frac{(1+\mu)\xi}{2\pi} \sum_{n=-\infty}^{\infty} \frac{f(\varepsilon_n)}{1+\mu f(\varepsilon_n)} , \quad \xi = 2\pi T/\Theta .$$

The results of numerical calculations of the function $\chi(\xi)$ for various values of the mass ratio M/m are shown in Fig. 1. At high temperatures, as can be seen from Eq. (2.9), the temperature shift of the Mössbauer line is given by the expression

$$\delta E = \frac{3E_0}{2mc^2} k_\beta T \tag{2.10}$$

which does not depend on the properties of the solvent crystal.

Let us now proceed to the treatment of the case $\mu \neq 0$, $\varkappa \neq 0$. Using the relation (2.1), we get the following system of integral equations which determine the Green's function $F^{\alpha\alpha'}(\tau, \tau')$,

$$F^{\alpha\alpha'}(\tau,\tau') - F_0^{\alpha\alpha'}(\tau,\tau') - \frac{\mu}{M}\int_0^\beta dz\, F_0^{\alpha\gamma}(\tau,z) F^{\gamma\alpha'}(z,\tau') - \\ -\varkappa\int_0^\beta dz\left[\mathcal{D}_0^{\gamma\alpha}(z,\tau) Y^{\gamma\alpha'}(z,\tau') + \left(Y_0^{\gamma\alpha}(z,\tau) - A(0,0)\mathcal{D}_0^{\gamma\alpha}(z,\tau)\right)\mathcal{D}^{\gamma\alpha'}(z,\tau')\right] \tag{2.11}$$

$$\mathcal{D}^{\alpha\alpha'}(\tau,\tau') = \mathcal{D}_0^{\alpha\alpha'}(\tau,\tau') - \frac{\mu}{M}\int_0^\beta dz\, \mathcal{D}_0^{\alpha\gamma}(\tau,z) F^{\gamma\alpha'}(z,\tau') - \\ -\varkappa\int_0^\beta dz\left[G_0^{\alpha\gamma}(\tau,z) Y^{\gamma\alpha'}(z,\tau') + \left(X_0^{\alpha\gamma}(\tau,z) - A(0,0)G_0^{\alpha\gamma}(\tau,z)\right)\mathcal{D}^{\gamma\alpha'}(z,\tau')\right]. \tag{2.12}$$

Here we have used the notations

$$\mathcal{D}^{\alpha\alpha'}(\tau,\tau') = \frac{\langle T u_0^\alpha(\tau) p_0^{\alpha'}(\tau') S(\beta)\rangle}{\langle S(\beta)\rangle}, \quad \mathcal{D}_0^{\alpha\alpha'}(\tau,\tau') = \overline{u_0^\alpha(\tau) p_0^{\alpha'}(\tau)}$$

$$G_0^{\alpha\alpha'}(\tau,\tau') = \overline{u_0^\alpha(\tau) u_0^{\alpha'}(\tau')}, \quad X_0^{\alpha\alpha'}(\tau,\tau') = \sum_l A^{\alpha\gamma}(0,l)\,\overline{u_l^\gamma(\tau) u_0^{\alpha'}(\tau')}$$

$$Y_0^{\alpha\alpha'}(\tau,\tau') = \sum_l A^{\alpha\gamma}(0,l)\,\overline{u_l^\gamma(\tau) p_0^{\alpha'}(\tau')}$$

$$Y^{\alpha\alpha'}(\tau,\tau') = \sum_l A^{\alpha\gamma}(0,l)\,\frac{\langle T u_l^\gamma(\tau) p_0^{\alpha'}(\tau) S(\beta)\rangle}{\langle S(\beta)\rangle}.$$

In the derivation of Eqs. (2.11) and (2.12) it has been assumed that the solvent crystal has cubic symmetry, so that $A^{\alpha\alpha'}(0, 0) = \delta_{\alpha\alpha'} A(0, 0)$ in a system of coordinates associated with the symmetry axes of the crystal.

For the model of a crystal with three degenerate branches of the spectrum of normal lattice vibrations we get as the equations for the Fourier transforms of the temperature-dependent Green's functions

$$\left(1 + \frac{\mu}{M} F_0(\varepsilon_n)\right) F(\varepsilon_n) = F_0(\varepsilon_n) - \varkappa\,\mathcal{D}_0(-\varepsilon_n) Y(\varepsilon_n) - \varkappa\,\bar{Y}(-\varepsilon_n)\mathcal{D}(\varepsilon_n) \tag{2.13}$$

$$\frac{\mu}{M}\mathcal{D}_0(\varepsilon_n) F(\varepsilon_n) = \mathcal{D}_0(\varepsilon_n) - \varkappa G_0(\varepsilon_n) Y(\varepsilon_n) - \left(1 - \varkappa X_0(\varepsilon_n)\right)\mathcal{D}(\varepsilon_n) \tag{2.14}$$

$F^{\alpha\alpha'}(\varepsilon_n) = \delta_{\alpha\alpha'} F(\varepsilon_n)$ and so on, and it is easy to determine the functions $D_0(\epsilon_n)$, $G_0(\epsilon_n)$, $\overline{X}_0(\epsilon_n)$, $\overline{Y}(\epsilon_n)$ if we recall that for the regular lattice

$$u_0^\alpha = \frac{1}{N^{1/2}} \sum_{q\sigma} \left(\frac{\hbar}{2M\omega_{q\sigma}}\right)^{1/2} V^\alpha(q\sigma)(b^+_{-q\sigma} + b_{q\sigma}) .$$

We easily get the following expressions for these functions.

$$\mathcal{D}_0(\varepsilon_n) = -\hbar \varepsilon_n \tau(\varepsilon_n) \quad , G_0(\varepsilon_n) = \frac{\hbar^2}{M} \tau(\varepsilon_n) \quad , \ F_0(\varepsilon_n) = M(1 - \varepsilon_n^2 \tau(\varepsilon_n))$$

$$\overline{X}_0(\varepsilon_n) = X_0(\varepsilon_n) - A(0,0) G_0(\varepsilon_n) = 1 - (\varepsilon_n^2 + \overline{\varepsilon^2})\tau(\varepsilon_n)$$

$$\overline{Y}_0(\varepsilon_n) = Y_0(\varepsilon_n) - A(0,0)\mathcal{D}_0(\varepsilon_n) = -\frac{M\varepsilon_n}{\hbar}\left[1 - (\varepsilon_n^2 - \overline{\varepsilon^2})\tau(\varepsilon_n)\right]$$

$$\tau(\varepsilon_n) = \frac{1}{N}\sum_q \frac{1}{\varepsilon_n^2 + \varepsilon_q^2} \quad , \quad \overline{\varepsilon^2} = \frac{1}{N}\sum_q \varepsilon_q^2 = \frac{\hbar^2 A(0,0)}{M} .$$

Eliminating $Y(\epsilon_n)$ from the equations (2.13) and (2.14), we get

$$\frac{\hbar}{M}(1+\mu) F(\varepsilon_n) = \hbar + \varepsilon_n \mathcal{D}(\varepsilon_n) . \tag{2.15}$$

On the other hand, we have the known relation /8/

$$-\varepsilon_n G(\varepsilon_n) - \frac{\hbar}{M}\mathcal{D}(\varepsilon_n)$$

$$G(\varepsilon_n) = \frac{\hbar^2}{M}\frac{1+\mu}{1+\varkappa}\tau(\varepsilon_n)\left\{(1+\mu)(1+\varkappa) - \left[\varkappa(1+\mu) + \mu + \frac{\varkappa}{1+\varkappa}\right]\varepsilon_n^2\tau(\varepsilon_n) - \varkappa(1+\mu)\overline{\varepsilon^2}\tau(\varepsilon_n)\right\}^{-1} . \tag{2.16}$$

For $F^{\alpha\alpha'}(\varepsilon_n)$ we find from Eqs. (2.15) and (2.16) the expression

$$F^{\alpha\alpha'}(\varepsilon_n) = \delta_{\alpha\alpha'} M \frac{(1+\varkappa)[1 - \varepsilon_n^2\tau(\varepsilon_n)] - \varkappa\overline{\varepsilon^2}\tau(\varepsilon_n)}{(1+\mu)(1+\varkappa) - \left[\varkappa(1+\mu) + \mu + \frac{\varkappa}{1+\varkappa}\right]\varepsilon_n^2\tau(\varepsilon_n) - \varkappa(1+\mu)\overline{\varepsilon^2}\tau(\varepsilon_n)}$$

Thus in this case we get as the expression for the temperature shift of the Mössbauer line

$$\delta E = -\frac{3E}{2mc^2}(1+\mu)\frac{1}{\beta}\sum_n \frac{(1+\varkappa)[1 - \varepsilon_n^2\tau(\varepsilon_n)] - \varkappa\overline{\varepsilon^2}\tau(\varepsilon_n)}{(1+\mu)(1-\varkappa) - \left[\varkappa(1+\mu) + \mu + \frac{\varkappa}{1+\varkappa}\right]\varepsilon_n^2\tau(\varepsilon_n) - \varkappa(1+\mu)\overline{\varepsilon^2}\tau(\varepsilon_n)} , \tag{2.17}$$

which goes over into the result already known if in Eq. (2.17) we set $\varkappa = 0$. At high temperatures δE does not depend on $\varkappa$.

In conclusion we express our gratitude to N. N. Matviishinaya for carrying out the numerical calculations in the Institute of Cybernetics of the Academy of Sciences of the Ukrainian SSR.

Institute of Physics, Academy of Sciences, Ukrainian SSR

REFERENCES

/1/ R. V. Pound and G. A. Rebka, Phys. Rev. Lett., 4, 274, 1960.

/2/ B. D. Josephson, Phys. Rev. Lett., 4, 341, 1960.

/3/ J. P. Schiffer, J. Heberle, P. Parks, Bull. Am. Phys. Soc. 6, 442, 1961.

/4/ A. A. Maradudin, P. A. Flinn, S. L. Ruby, Phys. Rev. 126, 9, 1962.

/5/ T. Matsubara, Progr. Theor. Phys. 14, 351, 1955.

/6/ A. A. Abrikosov, L. P. Gorkov, I. E. Dzyaloshinskii, ZhÉTF, 36, 900, 1959.

/7/ N. N. Bogolyubov and D. V. Shirkov, Introduction to the Theory of Quantized Fields (in Russian), GITTL, M. 1957.

/8/ I. P. Dzyub and A. F. Lubchenko, in this issue.

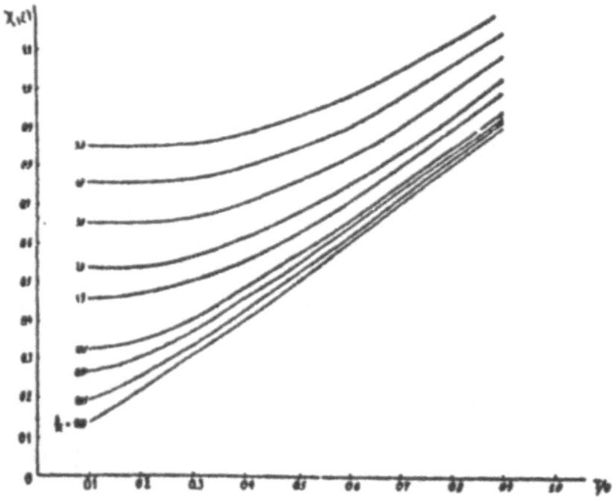

Fig. 1

Translated by W. H. Furry

THE MÖSSBAUER EFFECT AND THE FERROELECTRIC PROPERTIES OF IONIC CRYSTALS.

V. Dvorzhak, Ch. Muzikarzh, V. Yanovets

1. A characteristic property of ferroelectric crystals is that they show spontaneous polarization in some temperature range. The phase transition from the paraelectric state (in which the spontaneous polarization is equal to zero) to the ferroelectric state, which occurs at the definite temperature T_{tr} (transition temperature), is accompanied by a change in crystal symmetry, and by characteristic properties of the physical parameters.

The static dielectric constant ε_s varies in the paraelectric range according to the Curie--Weiss law

$$\varepsilon_s = \frac{4\pi C}{T - T_c} \tag{1}$$

where C and T_c are the Curie constant and the Curie temperature respectively. If the phase transition is a transition of the second kind, we have $T_{tr} \equiv T_c$. For transitions of the first kind, these temperatures are usually separated by several degrees, and $T_{tr} > T_c$. It is shown in /1,2/ that there is a relation between the ferroelectric transitions of ionic crystals and the anomalous properties of the crystal lattice vibrations. This relation follows from the equation (see /2/)

$$\frac{\varepsilon_s}{\varepsilon_\ell} = \prod_{\alpha=2}^{N} \frac{(\omega_\alpha^2)_L}{(\omega_\alpha^2)_T} \tag{2}$$

in which ε_1 is the dielectric constant measured at optical frequencies, and $(\omega_\alpha)_L$ and $(\omega_\alpha)_T$ are the longitudinal and transverse optical vibration frequencies for zero value of the wave vector $\vec{k}$. Further, N is the number of ions in the elementary crystal cell, and the subscript $\alpha = 1$ corresponds with the acoustical branch. The optical frequencies of nonferroelectric ionic crystals are in the infrared region. Following /1,2/ ferroelectric ionic crystals are characterized by the fact that the frequency of one of the transverse optical branches for $\vec{k} = 0$ takes on exceedingly low values even for $T = T_c$ equal to zero. In this case the dielectric constant would become infinite, and the crystal would be unstable to this anomalous branch. As a result, there is a change in crystal symmetry, the frequency of the anomalous branch increases, and spontaneous polarization appears. If it is assumed that the anomalous frequency ω_{an} depends on the temperature:

$$\omega_{an}^2 = \gamma (T - T_c), \tag{3}$$

while the temperature dependence is absent for the rest of the optical frequencies, the Curie--Weiss law (1) may be accounted for. (It is shown in /3/ for example that $BaTiO_3$ follows Eq. (3) quite well, if $\gamma = 1.00 \cdot 10^{22}$ $deg^{-1} sec^{-2}$).

2. Thinking of the ferroelectric transition in this way shows why it is important to make a more detailed study of the anomalous branch.

The properties of the lattice vibrations are determined by short range repulsion forces and Coulomb attraction forces. In nonferroelectric crystals, the repulsive forces are greater than the attractive forces, and the optical vibration frequencies occur in the infrared region of the spectrum. However, in crystals with appropriate symmetry, the Lorentz factors, which determine the magnitude of the local electric fields, are so large that the Coulomb forces reach values comparable with the repulsive forces. In these crystals, it is possible to have anomalous lowering of the frequency, and restoration of the ferroelectric state. As shown in /1/, the Lorentz factors decrease with increase in $|\vec{k}|$. It is accordingly to be expected that the attractive and repulsive forces are only comparable for small values of $|\vec{k}|$. As a result of this, there is only a small range of wave vectors $\vec{k} \sim 0$, in which the frequency takes on exceedingly low values. The remaining values of $\vec{k}$ correspond with frequencies in the infrared range. This is illustrated in /2/, where the simplest model of a ferroelectric substance is discussed, namely the NaCl crystal, in which the lattice parameters were changed in such a way that for $\vec{k} = 0$ one of the optical transverse frequencies took on a zero value. (See Fig. 1, in which LO and TO are the optical longitudinal and transverse branches, and LA and TA are the same thing for the acoustical branches).

In what follows, the true form of the anomalous optical branch will be replaced by the curve shown in Fig. 2, which retains the properties of the anomalous branch, at least in its general features. In accordance with the Curie--Weiss law, the low frequency part of the curve changes with temperature in the way shown in Eq. (3). The second part of the curve is assumed independent of the temperature.

3. We assume just now that a γ- transition (transition ener-

gy E_0, level width Γ suitable for observing the Mössbauer effect is possible in some one of the nuclei making up the elementary cell of the crystal. The Mössbauer line intensity is given by the formula $W(E) = W_0(E)\exp(-\sum_\alpha Z_\alpha)$, in which the subscript α is the number of the branch of the vibration spectrum of the crystal, and

$$Z_\alpha = \int d\omega\, \rho_\alpha(\omega)(R_\alpha/\hbar\omega)\left(n(\omega,T) + \tfrac{1}{2}\right). \qquad (4)$$

Here $n(\omega, T)$ is the mean number of phonons of energy $h\omega$, $\rho_\alpha(\omega)$ is the branch α phonon frequency distribution function, normalized to unity, and R_α is the part of the recoil energy of the nucleus (of mass M), going into the branch α:

$$R_\alpha = \left(E_0^2/2Mc^2\right)\left(r_\alpha/3\right)|\vec{e}_\alpha|^2 \qquad (5)$$

The polarization vectors of the vibrations, $\vec{e}_\alpha$, are normalized in such a way that $\sum_\alpha r_\alpha |\vec{e}_\alpha|^2 = 3$ and $r_\alpha = 1.2$ for the longitudinal and transverse vibrations respectively.

Consider the anomalous branch contribution ($\alpha \equiv$ an). For the curve shown in Fig. 2, we obtain

$$\rho_{an}(\omega) = \varkappa^2\delta(\omega - \omega_{an}) + (1 - \varkappa^2)\delta(\omega - \omega'_{an}). \qquad (6)$$

The parameter $\varkappa^2 = k_0/k_{max}$ gives the contribution made by the low frequency vibrations to the anomalous branch. (It may be assumed that $\varkappa^2 \ll 1$). Thus

$$Z_{an} = R_{an}\left[\frac{n(\omega'_{an},T)+\frac{1}{2}}{\hbar\omega'_{an}} + \varkappa^2\left(\frac{n(\omega_{an},T)+\frac{1}{2}}{\hbar\omega_{an}} - \frac{n(\omega'_{an},T)+\frac{1}{2}}{\hbar\omega'_{an}}\right)\right]. \quad (7)$$

For temperatures somewhat above T_c, the frequency ω_{an} is two orders of magnitude less than ω'_{an}, and the first term is greater than the second in the round brackets in Eq. (7). The presence of an anomalous branch ($\varkappa \neq 0$) shows up in the value of Z_{an} as compared with the case where no low frequency vibrations occur ($\varkappa = 0$). Thus, the anomalous branch weakens the Mössbauer line intensity.

4. The elementary cells of some ferroelectric materials already have nuclei in them which are suitable for observing the Mössbauer effect (for example, ferroelectric materials containing iron, Pb_2FeTaO_6 /4/, tungsten WO_3 /5/, etc.)[3]. It is also possible to introduce a suitable nucleus into the elementary cell as an impurity. What is then required is for the impurity not to change the anomalous properties of the lattice vibrations. It turns out /7/ that this is possible with single crystals of $BaTiO_3$ containing Co or Fe as impurities, if the impurity atoms are uniformly distributed in the angular points of the lattice, and that the ferroelectric properties of the crystal are retained even for concentrations of several percent.

Thus, measuring the temperature dependence of the Mössbauer effect at T_c would give some valuable information on the nature of the anomalous optical branch in various ferroelectric materials. More detailed calculations are being made.

3) See the detailed compilation of ferroelectric materials in /6/.

The authors thank the participants at the conference for their useful discussions.

Physics Institute, Academy of Sciences, CzechSSR, Prag, Czechoslovakia.

Physics and Mathematics Department, Karlov University, Prag, Czechoslovakia.

Translated by:Charles V. Larrick

REFERENCES

/1/ Anderson, P. W., Physics of Dielectrics. Transactions II of the All Union Conference, 1958 /Russian translation/ Academy of Sciences Press, USSR, Moscow, 1960

/2/ Cochran, W., Adv. Physics 9, 387, (1960)

/3/ Dvořak V., Janovec V., Czech. J. Phys. B 13 (1963)(in press)

/4/ Isupov, V. A., Agranovskaya, A. I., Khuchia N. P., Izv. AN SSSR, Physics series 24, (1960), 1271

/5/ Tanisaki S., J. Phys. Soc. Jap. 15, 566 (1960)

/6/ Fousek J. Čs. čas fys. 11, 495 (1960)

/7/ Arend H. T., Coufova P. Czech. J. Phys. B 12 (1962)(In press)

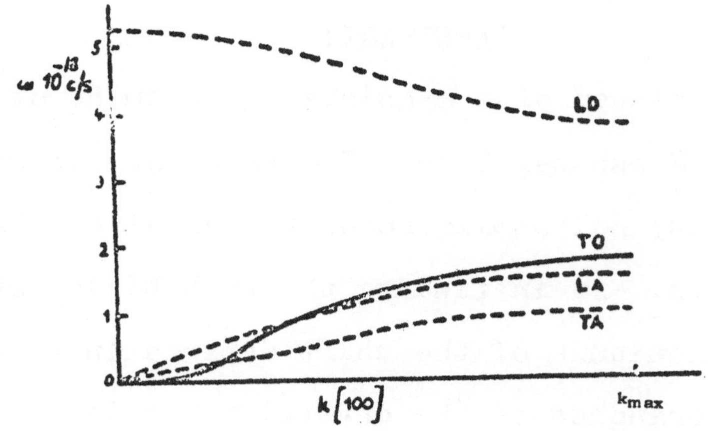

Fig. 1

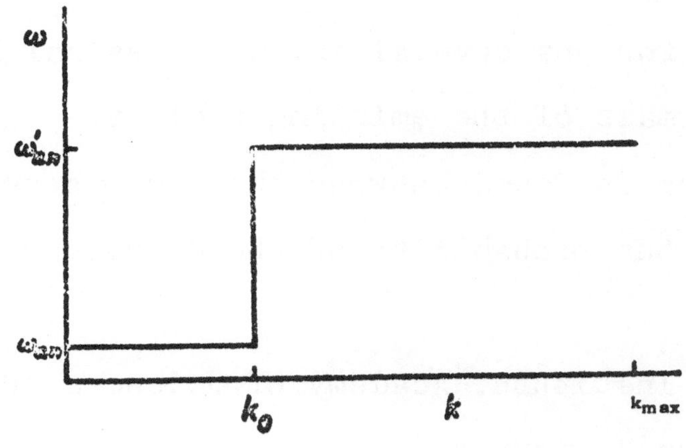

Fig. 2

THE THEORY OF THE THERMAL RED SHIFT OF THE MÖSSBAUER LINE

Yu. Kagan

(ABSTRACT)

Results are presented of a consistent treatment of the thermal red shift of the Mössbauer line. The value of the shift ΔE has been found for any polyatomic regular crystal for the whole range of temperatures. As in finding the probability of the Mössbauer effect, the magnitude of the shift depends in an essential way on the optical branches of the crystal vibration. A direct relation between ΔE and the heat capacity of the lattice occurs only for a monatomic crystal.

An expression is found for the shift in the case where the radiator is a nucleus foreign to the monatomic matrix. The temperature dependence is analyzed, particularly for the limiting cases $m'/m \ll 1$, and $m'/m \gg 1$.

It is shown that for any crystal in the classical limit, ΔE depends solely on the mass of the emitting nucleus.

A general relation is found between the temperature dependence of the line shift and the probability of the Mössbauer effect at $T \neq 0$.

Atomic Energy Institute, Academy of Sciences, USSR

Translated by: Charles V. Larrick

RECOILLESS γ EMISSION AND ABSORPTION BY ATOMS IN A MAGNETIC FIELD

M. V. Kazarnovskii, A. V. Stepanov

Recoilless γ emission and absorption by atoms (Mössbauer effect) is possible in those cases where there is no rigid relation between the energy and momentum of the atom, so that the atom can transfer momentum without changing its energy. So far, the Mössbauer effect has only been observed for γ-transitions of atomic nuclei in condensed systems (crystals, polymers). However, other mechanisms of recoilless γ emission and absorption are in principle possible. This occurs, first, when the quanta are emitted or absorbed by an ionized atom in a strong magnetic field, with the direction of propagation of the quantum perpendicular to the lines of magnetic force*.

The probability W of this process in a uniform magnetic field may be calculated. If it is assumed for purposes of calculation that the atoms in the magnetic field have a Boltzmann distribution with temperature T, the following expression is obtained for W:

$$W \sim e^{-a\frac{1+s}{1-s}} I_0\left(2\,\frac{\sqrt{s}}{1-s}\right) \tag{1}$$

*It is easily verified that the angle between the direction of propagation of the γ- quantum and the magnetic field will not depart from $\pi/2$ by more than $(2Mc^2\Gamma)^{\frac{1}{2}} \approx 10^{-2}$--$10^{-4}$ (E is the energy, Γ is the level width, and M is the mass of the nucleus).

$$a = \frac{E^2}{2Mc^2\hbar\omega} \; ; \quad s = e^{-\frac{\hbar\omega}{kT}}, \qquad \omega = \frac{eHz}{Mc} \qquad (1)$$

where Z is the ionization multiplicity factor of the atom, and $I_0(x)$ is the Bessel function of imaginary argument. Practically always, $\hbar\omega/kT \ll 1$, and $\underline{a} \gg 1$, so that I_0 may be replaced by its asymptotic expression. Then

$$W \backsim \frac{\hbar H e z}{E\sqrt{2\pi kTM}} e^{-\frac{E^2}{8kTMc^2}}$$

Note, that the exponent of the exponential is nearly zero for sufficiently soft γ's. Finally

$$W \backsim 0{,}83 \cdot 10^{-5} \frac{Hz}{E\sqrt{MT}}$$

where H is in oersteds, E is in electron volts, T is in $^{\circ}$K, and M is in atomic mass units. I.e., the probability of recoilless γ emission and absorption by ionized atoms in a magnetic field is extremely small, and experiments of this sort can scarcely be done directly.

A second method of recoilless γ emission and absorption is possible only for atoms (and nuclei) having a non-zero magnetic moment. In a magnetic field, these atoms and nuclei have different (but discrete) values of energy for a given momentum. For example, in a very strong magnetic field, where the coupling between the nuclear magnetic moment μ_0 and the electron shell may be neglected, the nucleus with spin I has $2I+1$ energy values for a given momentum, the energies being equidistantly located with the spacing $\mu_0 H/\sqrt{I(I+1)}$, i.e., if the momentum transferred is such that

$$\frac{E^2}{2Mc^2} = \frac{m\mu_0 H}{\sqrt{J(J+1)}} \qquad (m \leqslant 2J) \qquad (4)$$

the equation being satisfied with an accuracy of the order of $2\Gamma Mc^2/E^2 \sim 10^{-4}$--$10^{-2}$ (the relative spread in the emission energy from the natural width of the γ-line), it is then possible to have emission and absorption of γ- quanta with an energy exactly equal to the energy of the excited nucleus (in a number of cases also greater by the factor $\mu_0 H/\sqrt{I(I+1)}$). The energy transferred with the momentum is made up for by a change in energy of the nucleus in the magnetic field.

However, even in this simple case, as is easily seen from Eq. (4), the effect can evidently only be observed experimentally if nuclei are found with the first level at $E \lesssim 100$ eV, and in addition with a quite small internal conversion coefficient. It is more realistic to perform the experiment in a "weak" field, where the coupling between the nuclear magnetic moment and the electron shell is not destroyed, so that the γ- emission can be accompanied by a change in the projection of the spin of the whole atom on the direction of the magnetic field.

If the distance between the hyperfine structure levels, ΔE_{hf}, were comparable or greater than $E^2/2Mc^2$, Eq. (4) would again hold, but with the difference that the magnetic moment of the atom must be substituted for the nuclear magnetic moment μ_0, i.e., a quantity 10^3 times greater, and further the spin of the atom must be substituted for the nuclear spin I. Further, if the spread in the initial kinetic energies of the atom, ΔE, is less than Γ, the transition probabilities are simply given in terms of the Klebsha--Gordon coefficient. In the opposite case it is a fraction of the order

$$\left(\frac{2 \Delta E_{hf}}{E_a^2 / Mc^2} \right)^2 \frac{\Gamma}{\Delta E} .$$

P. N. Lebedev Physics Institute, Academy of Sciences, USSR

Translated by: Charles V. Larrick

ANGULAR CORRELATION IN THE MÖSSBAUER EFFECT

A. Gel'berg

The angular correlation of cascade γ's is one of the most widely used methods of determining the spins of nuclear states. In resonance scattering of γ-radiation, investigating angular correlation reduces to finding the angular distribution of the scattered quanta about the primary direction.

If polarization is not observed, the distribution function is of the form /1,2,3/

$$W(\theta) = \sum_{K} \sum_{LL'} F_K(LL'I_0I_1) \sum_{LL'} F_K(LL'I_0I_1)\, P_K(\cos\theta) \quad (1)$$

where θ is the scattering angle, $\underline{k}$ is an even number, I_0 and I_1 are the spins of the ground and excited states respectively, and L and L' give the multipolarity of the radiation. Values of F_k are found in the tables in /3/.

Experiments of this sort have been made by Metzger /4,5/, using ordinary resonance scattering. In the Mössbauer effect, conditions are more favorable, because of the large absorption cross-section. The possibility of using recoilless scattering for this purpose was pointed out in the first Mössbauer conference /6/.

The following three cases may be distinguished:

1. The conversion coefficient is small, as for example in Eu^{153}, Tm^{169}, Gd^{154}, etc. In this case, the angular distribution (1) may be observed directly. The success of the experiment de-

pends solely on the effective cross-section of the absorber, and the absorption in the scatterer, which sets an upper limit to the scatterer thickness.

2. If, as a result of the small energy of the transition, large value of Z, or higher multipolarity, the internal conversion coefficient is large, observing the scattered γ- radiation becomes complicated. If, for example, the absorption in an absorber containing Fe^{57} is 10%, and the value of the conversion coefficient is $\alpha = 14$, the scattered radiation intensity is of the order of 1%. Under the conditions in which angular distribution is usually observed, the relative intensity that the detector can record may get as low as $\sim 10^{-5}$.

If the energy of the transition is not too small, the angular distribution of the conversion electrons may be observed.

The angular correlation function is of the form (1) with one of the F_k (for each <u>k</u>) replaced by $b_k F_k$, where the "particle parameter" b_k was calculated by Biedenharn and Rose /2/.

These experiments are limited by the source thickness. If Bothe's /7/ approximate formula is used for Ag^{109}, the mean multiple scattering angle of the K-conversion electrons (52.5 keV) in a 0.2 mg/cm^2 thick absorber is $\sim 30^{\circ}$.

3. The above possibility is absent in case the energy of the transition is so low as to make the source thickness too small. In this case, a study may be made of the angular distribution of the X-rays emitted after internal conversion. Dolginov /8/ has calculated the γ-X angular correlation. If the circular polarization of the γ's is not measured, the correlation is only observed for transitions to

the sub-level L_{111} (2 P 3/2).

Here the angular correlation function is of the form

$$W(\theta) = 1 + AR_{I_1 I_0} P^{I}_{20} P_2 (\cos\theta) \; , \qquad (2)$$

where A is a constant depending on the moment of the electron incident at the L_{111} level, P^{I}_{20} is a coefficient depending on I_0, I_1, L, and L'. $R_{I_1 I_0}$ depends on the same quantum numbers, and on the electron radial integrals $R_1, \ldots, R_6$, found in the paper by Rose /2/. If L = L' (pure γ- transition), $R_{I_1 I_0}$ does not contain the nuclear matrix elements.

This phenomenon may be observed under the following conditions:

a) no K-conversion because of the low energy of the transition.

b) large conversion coefficient, and, at the same time, large relative intensity of the L_{111} conversion line.

The table given below contains several parameters related to the measurements

Table 1

Nucleus	Conversion Coefficient	Energy of Transition (keV)	Relative Conversion Line Intensities L_1	L_{11}	L_{111}
Dy^{161}	2.1	25.7	1.0	0.8	1.1
Au^{197}	2.5	77	1.0	0.46	0.34
Ir^{193}	11	73	0.03	1	0.97*

*Theoretical value.

Unfortunately, it is at present impossible to make a complete quantitative interpretation of the experiment, since there are no tables of the radial integrals R_1 for the L sub-levels.

As far as observing the X-rays emitted after conversion are concerned, this has been demonstrated by Frauenfelder's experiment /10/.

Atomic Physics Institute, Bucharest

Translated by: Charles V. Larrick

REFERENCES

/1/ S. Devons, L. J. B. Golfarb, Angular Correlations (Hdb. of Physics. Vol. XLII, Berlin, 1957)

/2/ L. C. Biedenharn, M. E. Rose, Rev. Mod. Phys. 23, 729 (1953)

/3/ A. H. Wapstra, G. J. Nijgh, R. van Lieshout, Nuclear Spectroscopy Tables, Amsterdam, 1959

/4/ F. Metzger, Phys. Rev. 98, 200 (1955)

/5/ F. Metzger, Phys. Rev. 101, 286 (1956)

/6/ Mossbauer Effect, University of Illinois, June 1960, edited by F. Frauenfelder and H. Lustig

/7/ W. Bothe, Handb. der Physik, Bd. 22/2, Berlin (1933)

/8/ A. Z. Dolginov, ZhÉTF, 34, 931,(1958)

/9/ M. E. Rose, Multipole Fields (New York, 1955)

/10/ H. Frauenfelder, D. B. F. Cochran, D. E. Nagle, R. D. Taylor Nuovo Cim. 19, 183 (1961)

INVESTIGATIONS OF THE MÖSSBAUER EFFECT IN Sn AND Fe COMPOUNDS

ISOMERIC CHEMICAL SHIFTS AND QUADRUPOLE SPLITTING OF γ-LINES

V. S. Shpinel'

Investigations of the resonance absorption of γ-rays by nuclei of solids have revealed the superfine structure and energy shifts of γ-lines induced by the interaction of nuclei with the electric and magnetic fields created by charges outside the nucleus.

In this paper we shall limit ourselves to effects related mainly to electrostatic interactions. In this case the Hamiltonian of the interaction of the nucleus can be written in the form:

$$H = V + Q \nabla E \qquad (1)$$

The energy of Coulomb's interaction V depends on the distribution of the charge density in the nucleus $\rho(r)$ and on the density of electrons in the region of the nucleus $\psi^2(o)$

$$V = -e \int \psi^2(o) \varphi \, dv, \qquad (2)$$

where the potential of the nucleus is

$$\varphi = \int \frac{\rho(r')dr'}{|r - r'|} \, . \qquad (3)$$

The second term in (1) gives the interaction of the quadrupole moment of the nucleus Q with the tensor of the gradient of the electric field ∇E. In the general case the Hamiltonian of this interaction has the form

$$H_Q = \frac{eQ}{2I(2I - 1)} \sum_{i,j} q_{ij} I_i I_j \quad (i, j = 1,2,3), \qquad (4)$$

where q_{ij} are the components of the tensor of the gradient of the electric field

$$q_{ij} = \frac{\partial^2 V}{\partial x_i \partial x_j} ,$$

and I is the operator of the spin of the nucleus.

These interactions as well as the magnetic interaction, which is omitted in (1), lead to a number of effects which are manifest in optical spectra of the fine and superfine structures and in isotopic and isomeric shifts of spectral lines in the range of resonance radio frequencies. The effect of these interactions on the form of the Mössbauer spectra provides additional information concerning the nucleus and, in particular, the medium surrounding the nucleus.

Isomeric Chemical Shift

Kistner and Sunyar /1/ found that the energies of the γ-lines shift according to the compounds in which the Mössbauer nucleus is located. They observed a small shift of the center of gravity of the Zeeman absorption spectrum in Fe_2O_3, using Co^{57} incorporated in stainless steel as the source.

Much larger shifts were observed by Bryukhanov, Delyagin, Zvenglinskii, and Shpinel' /2/ in the case of the 23.8 keV γ-line of Sn^{119} in different tin compounds. Figure 1 illustrates the shift of single lines in the resonance absorption spectra of a number of tin compounds obtained with SnO_2 as the source.

The displacement of the line δ with respect to zero can be determined quite precisely in such spectra. The difference in the energies of the emission and absorption lines can be obtained from Eq. (2). The energy added to γ-quanta by the Coulomb interaction results from the change in this interaction ΔV due to the change in the charge distribution of the nucleus $\Delta\rho$

$$\Delta V = -e \int \psi^2(o)\, \Delta\rho \, dv, \qquad (5)$$

The observed spectral shift δ, which is nothing but the change in the value of ΔV for the source and the absorber or for different absorbers, results from the change in the probability density $\psi^2(o)$ in them.

$$\delta = e \int \left[\psi_1^2(o) - \psi_2^2(o)\right] \Delta\rho \, dv \qquad (6)$$

This equation shows that if the electron densities on the nucleus are known then the information concerning the change of the charge distribution in the nucleus resulting from the transition of the nucleus into the excited state can be obtained by measuring the values of δ .

The electron density on the nucleus, $\psi^2(o)$, is determined essentially by the s-electrons. The electron density can be determined experimentally from the splitting of the fine and superfine structures of optical spectra or from Knight shifts, for example, or can be calculated if the precise wave functions of s-electrons are known. If these data are lacking one can use the approximate Fermi-Segrè formula:

$$\psi_n^2(o) = \frac{Zz^2}{\pi a_n^3 n_o^3} \left(1 - \frac{d\Delta}{dn}\right) \qquad (7)$$

where $\psi_n^2(o)$ is the density probability on the nucleus for s-electrons having the principal quantum number n; Z is the charge of the nucleus; z is the charge of the ion in whose field the electron moves; n_o is the effective quantum number of the nth stationary state; and $\sigma = n - n_o$ is the quantum defect of the nth state.

This equation was derived for alkali metals, but, as Foldy /3/ has shown, it can be applied to other cases provided σ is a smooth function of n.

Thus, when $\psi^2(o)$ is known and when the change in the charge distribution $\Delta\rho$ can be calculated on the basis of a given model of the nucleus, the observed values of δ make it possible to check the validity of the model used.

If the charge distribution in the nucleus is characterized by some equivalent uniform distribution, with radius R given by the relationship

$$\frac{3}{5} ZR^2 = \langle r^2 \rangle \tag{8}$$

then one obtains the following expression for

$$\delta = c \left[\psi_1^2(o) - \psi_2^2(o) \right] \frac{\delta R}{R} \tag{9}$$

Here δR is the change in the radius of the equivalent uniform charge distribution in the ground and in the excited states, and c is the value calculated in /4, 5/, for example.

The considerable number of experimental data on isomeric chemical shifts which have been accumulated lead to certain conclusions.

Thus, De Benedetti, Lang, and Ingalls /6/, using bi-valent and tri-valent iron compounds, showed that the isomeric chemical shifts for iron compounds with different valencies are very different from each other and are relatively independent of the elements combined with iron. Using the difference in the shifts for bi-valent and tri-valent iron ($\delta_{++} - \delta_{+++}$) and the theoretical values of Watson's wave functions for free iron ions $[\psi_{++}^2(o) - \psi_{+++}^2(o)]$, they found that $\delta R/R = -10^{-3}$ ($R_g > Re$).

The effect of the crystalline environment on the electron density at zero was neglected in this calculation.

Walker, Wertheim, and Jaccarino /7/ made a systematic investigation of the isomeric chemical shift for bi-valent and tri-valent iron compounds incorporated into metals of the d-group (Fe, Co, Ni, Mn, Cr, Mo). The interpretation of the shifts given by these authors is shown in Fig. 2. The y axis represents the total density of s-electrons in atomic units. The values of $2\sum_{n=1}^{3}\psi_n^2(0)$ calculated by Watson for different $3d^n$ configurations of iron between n = 4 and n = 8 are indicated on the scale to the left. The scale on the right for the observed shifts of the lines is graduated in terms of the density of s-electrons so that the difference in the electron densities for the compounds with the strongest ionic character, $Fe^{2+}(FeF_2)$ and $Fe^{3+}(Fe_2(SO_4)_3 \cdot 6\,H_2O)$, corresponds to the difference in the shifts for these compounds. The solid lines represent the density of s-electrons for the assumed $3d^n 4s^x$ configuration. They are given here on the assumption that the density for such a configuration has the form $|\psi(3d^n)|^2 + X|\psi_{4s}(o)|^2$, where $|\Psi_{4s}(o)|^2$ is calculated by the Fermi-Segrè formula and X is the degree of filling of the states (along the x axis). The experimental points of this diagram make it possible to identify the configurations of the electrons for iron in different states. These authors found that the variation of the radius of the charge of the nucleus is $\delta R/R = -1.8 \cdot 10^{-3}$.

Thus, it can be considered proved that the charge distribution radius R for the nucleus of Fe^{57} in the excited state is smaller than for the nucleus in the ground state, and the difference is quite significant. It should be noted that the change of the charge radius for A = 57 resulting from the addition of one particle is $\delta R/R \approx 1/3(\delta A/A) = 5.9 \cdot 10^{-3}$.

Investigations of the 23.8 keV γ-line of Sn^{119} at Moscow State University showed that the value of the isomeric chemical shift for different tin compounds also depends on the valency of the compound /8/. This result can be understood if one assumes that out of the four outer electrons of the tin atoms with the $5s^2 5p^2$ configuration only two p-electrons participate in bi-valent compounds. These electrons affect the electron density at zero much less than the s-electrons. In quadri-valent compounds two more s-electrons enter the chemical bond, and this leads to a sharp decrease of the isomeric chemical shift. We may conclude, therefore, that $\Psi^2(o)$ is smaller for quadri-valent tin than for bi-valent tin $[\Psi_{IV}^2(o) < \Psi_{II}^2(o)]$. Since the observed difference in shifts for the bi-valent (II) and the quadri-valent (IV) compounds is $\delta_{II} - \delta_{IV} > 0$, it follows from (9) that $\delta R > 0$, i.e., the charge distribution radius for the nucleus is greater in the excited state than in the ground state. Boyle, Bunbury, and Edwards /9/, who investigated the isomeric chemical shift of bi-valent and quadri-valent tin compounds, came to the same conclusions. They calculated $\psi^2(o)$ by Eq. (7) and the difference between the values of $\psi^2(o)$ for bi-valent and quadri-valent tin by the Crawford and Schawlow method /10/, in which screening is taken into account. They obtained $\delta R/R = +1.1 \cdot 10^{-4}$ for the 3/2+ excited state. These authors consider that the precision of their calculations is better than 30%. In principle, the isomeric chemical shift makes it possible to obtain more detailed information concerning the nucleus than that yielded by the mean square approximation of the radius described above. In fact, one can calculate the isomeric chemical shift if one chooses the wave functions for the states of the nucleus to be expected from a given model and one can determine how well these functions approximate the states considered. Shirley /11/ made such calculations for the 77 keV γ-transition in Au^{197}, where the number of protons in the nucleus is odd. In the case of a one-particle model one can assume that excitation is due to the Coulomb's interaction with the odd proton or hole. This proton (hole) is described by the shell wave function of $3S_{\frac{1}{2}}$ (excited state) and $2d_{3/2}$ (ground state). The interaction with the proton subshells of the two states does not affect the results. However, one must assume that the wave function of the 6s-electrons of the free gold atoms is important in this interaction. In the vicinity of the nucleus this wave function must be identical to that of metallic gold. Radial wave functions of the nucleus were calculated for three different types of potential: the rectangular hole potential, harmonic oscillator potential, and the Wood-Saxon potential. It turned out that the value of the isomeric chemical shift is very sensitive to the type of nuclear potential.

Comparison of the calculated shift and the shift found experimentally in the case of gold added to platinum, iron, cobalt, and nickel shows that the best agreement between these results comes from the use of the Wood-Saxon potential, which makes it possible to choose the configuration of protons.

However, these considerations do not apply to nuclei with an odd neutron (Fe^{57} and Sn^{119}, for example) because the one-particle neutron transition in this model should not lead to a shift at all. The fact that the shift does occur for nuclei such as Fe^{57} and Sn^{119} makes us assume that the excitation of the neutron has a polarizing effect on the proton center of the nucleus, inducing deformation and a change in the charge distribution.

Bryukhanov, Delyagin, Opalenko, and Shpinel' /12/ investigated resonance absorption spectra for different tin compounds. They found a number of relationships which are apparently quite important for the determination of the character of the chemical bond. It is clear that the change of electron density which affects the shift depends essentially on the change in the configuration of valence electrons responsible for the chemical bond. Thus, it was found that the shifts observed in the case of quadri-valent tin compounds depend on the electronegativity of the element combined with tin and that the shift increases when the electronegativity of the element decreases in the order of the electronegativity series: F, O, Cl, Br, Sn, I (see Fig. 3 in the report by Aleksandrov, Delyagin, et al.).

The more electronegative the element the more the electron charge is attracted, and therefore the electron charge density on the tin nucleus decreases. The decrease of the charge density on the tin nucleus leads to a decrease of the shift, as we have shown in the case of bi-valent and quadri-valent compounds. For Fe^{57} the situation is reversed, the decrease of electron density $\psi^2(o)$ leading to an increase in the shift, as one would expect from the opposite sign of the radius R of the nucleus.

Quadrupole Splitting

In some cases the absorption spectra for Sn^{119} observed by a group of researchers at Moscow State University had a doublet structure which was interpreted as the result of quadrupole splitting of the 3/2+ excited level. These spectra were observed for the first time in the case of absorbers made of SnO and SnF_2. These spectra are shown in Fig. 3. One can see that the spectrum consists of two components of equal intensities and that the center of gravity of the spectrum is displaced with respect to zero velocity. This displacement is due to the isomeric chemical shift with respect to the emission line of the source SnO_2 used in these experiments. The measurements were made at room temperature and at the temperature of liquid nitrogen, but no change in the value of splitting Δ was noted within the limits of the experimental error. In the case of a field of axial symmetry the value 2Δ is equal to the quadrupole bond constant $2\Delta = eQq$. If the value of quadrupole interaction is small with respect to the natural width of the line then the line widens instead of splitting. To obtain the constants eQq in this case it is necessary to make measurements with absorbers of different thicknesses and treat the experimental spectra correspondingly. Bykov and Pham Zuy Hien /13/ made the most complete calculation of parameters of Mössbauer spectra, which was presented at this meeting. In our early experiments with white metallic tin (β-Sn) we observed single lines at room temperature and at the temperature of dry ice, while at the temperature of liquid nitrogen the lines were double. But experiments repeated with thin absorbers did not confirm the doublet structure of the lines in β-Sn at the temperature of liquid nitrogen.

The splitting of the lines in the case of SnO and SnF_2 absorbers was confirmed in /9/. Single lines for β-Sn observed in /14/ cannot be taken as proof of the absence of quadrupole splitting, since thick absorbers were used in this investigation. Absorption spectra obtained with single crystals of β-Sn /15/, which will be presented at this meeting, indicate the presence of a quadrupole interaction which depends on the temperature.

Measurements with single crystals are of great importance. In this case the intensity of the components of the doublet must depend on the angle between the direction of the incident γ-quanta and the C axis of the crystal. This makes it possible to determine the sign of the interaction constant and also the asymmetry parameter $\eta = \frac{q_{xx} - q_{yy}}{q_{zz}}$.

Investigations of different tin compounds carried out at Moscow State University showed that: 1) quadrupole splitting occurs only in bi-valent tin compounds, and 2) the value of the quadrupole interaction increases regularly with increasing electronegativity of atoms combined with tin. The spectra of a few bi-valent tin compounds are shown in Fig. 3, where it can be seen that the greatest splitting occurs in fluoride compounds, where the electronegativity is greatest. Starting with $SnBr_2$ the lines cease to split. In quadri-valent Sn compounds the absorption line is not split (Fig. 1).

These relationships can easily be understood on the basis of a simple model of chemical bonds between tin atoms. In these compounds the tin atom forms four equivalent bonds which are responsible for the symmetrical (tetrahedral) surrounding of the tin atom. This symmetrical environment cannot create any significant field gradient on the tin nucleus. On the other hand, in bi-valent tin compounds the two outer p-electrons must create directed p-bonds, which results in an asymmetrical molecule. Large field gradients are created in such asymmetrical molecules. In our discussion we have neglected the effect of the crystalline environment. Generally speaking, field gradients created on the nucleus depend, first of all, on the configuration of the electrons of the atom. The charges of the molecule itself and the contribution of the crystalline field have much less effect. These factors must be taken into account when the internal atomic field is small. The external crystalline field q_k can be accounted for by the Sternheimer anti-screening factor

$$q = (1 - \gamma_\infty) \sum \left(\frac{3 \cos^2\theta_i - 1}{r_i^2}\right) e_i \equiv (1 - \gamma_\infty) q_k \tag{10}$$

where γ_∞ is the antiscreening factor /16/, r_i is the distance from the ith charge, θ_i is the angle between $\vec{r}_i$ and the principal axis of the tensor of the field gradient, e_i is the charge of the ith ion; the summation is over all ions in the lattice with the exception of one corresponding to r = 0.

Experiments with organometallic tin compounds made at Moscow State University together with the Institute of Chemical Physics and the Institute of Petrochemical Synthesis, Academy of Sciences, USSR, give results in agreement with the assumption that intramolecular fields play the principal role. We compared the spectra of given molecular compounds having different microstructures, e.g., the spectra of polymers and the corresponding monomers containing Sn in the case of pure compounds and solid solutions of these compounds. The spectra were identical in all cases. In cases where splitting of lines was observed the magnitude of the splitting was the same.

De Benedetti et al. /6/ investigated absorption spectra in bi- and tri-valent iron compounds. They showed that the magnitude of quadrupole splitting depends on the valency of the compound.

The magnitude of the splitting was great only in the case of bi-valent compounds. No splitting occurred in tri-valent iron compounds. Kistner and Sunyar /1/ found indications of quadrupole splitting in tri-valent Fe_2O_3, but the constant eQq turned out to be incorrect. Buchannon and Wertheim remeasured this constant in Fe_2O_3 and obtained eQq/h = 9.0 MHz /17/.

The magnitude of the splitting in the case of bi-valent iron compounds and its weak dependence on the temperature indicate that splitting is affected mainly by intraatomic fields. The weak splitting of tri-valent iron compounds results from the fact that the configuration of the external electrons in the Fe^{3+} ion is $3d^5$, i.e., the d-shell is half filled, and therefore its electrons cannot induce a field gradient. In this case the maximum gradient results from the closest ions. In this case the value of q can be calculated by Eq. (10).

Alff and Wertheim /18/ measured quadrupole splitting in the case of Fe^{57} situated at the octrahedral (a) and tetrahedral (d) sites of the $Y_3Fe_2(FeO_4)_3$ compound. The values of eQq/h were -22 and -18 MHz respectively. From these figures and the value of the γ_∞-factor for Fe^{3+} determined in /19/ Burns calculated the sum over the whole lattice and obtained for the quadrupole moment of the excited state of Fe^{57}

$$Q \simeq +0.4 \cdot 10^{-24}\ \text{cm}^2. \qquad (11)$$

The quadrupole moment of the 3/2+ state for the Sn^{119} nucleus is more difficult to determine because the molecular wave functions are unknown for tin compounds in which quadrupole splitting occurs. The field gradient on the nucleus can be calculated approximately by the method proposed by Tounes and Dailey /20/. The three factors contributing to the field gradient on the nucleus of the molecule can be examined separately: 1) gradient due to valence electrons for which the probability of being in the vicinity of the nucleus is great, 2) the effect of the electrons of the inner shell bound to the nucleus, and 3) the effect of the other charges in the molecule, such as electrons and ions separated from the nucleus by a distance of the atomic radius or greater. It was shown that the valence electron is responsible for the major contribution to the value of q and that the neglect of the other two factors leads to an error not exceeding 10% /20/.

In the case of an atom with one valence electron outside the closed shell the wave function of this electron in the molecule can be developed into atomic wave functions ψ_{nlm}

$$\psi = \sum_{nlm} a_{nlm} \psi_{nlm} \qquad (12)$$

and the contribution from this valence electron to the field gradient is found by the expression

$$\frac{\partial^2 V}{\partial r^2} = e \int \psi^* \left[(3 \cos^2\theta - 1)/r^3 \right] \psi \, d\tau \qquad (13)$$

where θ is the azimuthal angle with respect to the binding axis which is the quantizing axis. Using this reasoning, Boyle et al. /9/ calculated the gradient of the field on the tin nucleus in SnO and SnS compounds for which the eQq/h constants were measured. The problem is reduced essentially to determining the gradient of the field created by one of the 5p-electrons and to calculating the number of these electrons (in other words, reduced to the calculation of the value of a^2_{510}). They obtained for the quadrupole moment of Sn^{119} in the 3/2+ state

$$Q = -8 \cdot 10^{-26}\ cm^2.$$

The authors could not find the origin of the error in this calculation, but they consider that the precision of this value is 50%. The sign of the quadrupole moment was determined from measurements of the spectra of tetragonal SnO. In these measurements the absorber (SnO powder) turned out for unknown reasons to be a single crystal (the authors indicating that this was happy chance). In the case of a single crystal the components of the doublet have different intensities, and from the difference one can determine the sign of the constant of the quadrupole bond. The sign turned out to be positive. The gradient of the field $\partial^2 V/\partial Z^2$ must be negative, since it is created by the p_z orbit (m = 0), which for SnO and SnS is the nonbinding orbit determining the gradient of the field. Alekseevskii et al. /15/ will present to this meeting a report on the character of the spectra of quadrupole splitting in the case of single crystal absorbers.

The application of the Mössbauer effect to the investigation of organic compounds is very interesting. The possibility of such investigations was demonstrated by the study of organometallic compounds of Sn made at Moscow State University in collaboration with the Institute of Chemical Physics, Academy of Sciences, USSR /21/. The molecules of these compounds consist of a great number of atoms, and their composition can be varied within wide limits. It is also possible to vary the field acting on the nucleus. For example, in our first investigations we obtained absorption spectra for compounds of the SnR_iX_{4-i} type (i = 0, 1, 2, 3, 4), where R is the organic radical and X is a halogene. The spectra obtained in the case where R is the phenyl group and X is a Cl atom are shown in Fig. 4. One can see that the line is a singlet when an Sn atom is surrounded by four phenyl groups. This is in agreement with the assumption that in such a molecule four equivalent bonds are formed as the result of sp^3 hybridization, and this leads to a tetrahedral (symmetrical) environment of the Sn atom, and there is either none at all or a very weak field gradient on the Sn nucleus. If one of the phenyl groups is replaced with a Cl atom the singlet is split into a doublet and the center of gravity of the whole spectrum shifts. Further replacement of phenyl groups with Cl atoms increses the splitting and the shift. Finally, when the Sn atom is surrounded by four Cl atoms there is again a singlet. The splitting into a doublet can also be interpreted in terms of the quadrupole interaction. But it should be noted that the components of the doublets are not equal in the case of $Sn(C_6H_5)_2Cl_2$ compounds. This fact requires special explanation. The different intensities of the components of the doublets and their different widths are also found in other organometallic tin compounds. One can advance different hypotheses to explain the observed form of the spectrum, for example, the presence of a weak magnetic field in the molecule. One cannot exclude the possibility of the existence of two different isomeric shifts, although this does not seem very probable.

A great variety of organometallic tin compounds have been investigated at Moscow State University in collaboration with the Institute of Petrochemical Synthesis, Academy of Sciences, USSR, and a certain number of relationships were established. Some of the results of these investigations will be reported at this meeting.

We hope that these investigations will provide more complete information on electric fields acting on the Sn nucleus and thus make it possible to obtain more precise data on the quadrupole moment and the charge distribution in the case of the Sn nucleus in the 3/2+ excited state.

These investigations show that we are dealing with an important new method of investigating the structure of organic molecules. It is very important that these studies be extended to iron-organic compounds using a Co^{57} source. The investigation of iron-organic compounds has already begun abroad.

It should be noted that the nuclear quadrupole interaction can be studied by the method of radiofrequency quadrupole resonance. With this method it is possible to measure the constants of the quadrupole bonds of the ground states of nuclei with great precision, but it cannot be applied to cases where the spin is less than 1. In particular, this method cannot be used for nuclei such as Sn^{119} and Fe^{57}, since the spin $I = \frac{1}{2}$ in the ground state.

Institute of Nuclear Physics, Moscow State University

REFERENCES

/1/ O. C. Kistner, A. W. Sunyar, Phys. Rev. Lett., 4, 412 (1960).

/2/ V. A. Bryukhanov, N. N. Delyagin, V. Zvenglinskii, V. S. Shpinel', ZhÉTF, 40, 713 (1961).

/3/ L. L. Foldy, Phys. Rev., , 1093 (1958).

/4/ A. R. Bodmer, Nucl. Phys., 21, 347 (1961).

/5/ Ya. Smorodinskii, Journ. of Phys. USSR, 10, 419 (1946).

/6/ S. De Benedetti, G. Lang, R. Ingalls, Phys. Rev. Lett., 6, 60 (1961).

/7/ L. R. Walker, G. K. Wertheim, and V. Jaccarino, Phys. Rev. Lett. 6, No. 3, 98 (1961).

/8/ V. S. Shpinel', V. A. Bryukhanov, N. N. Delyagin, ZhÉTF, 41, 1767 (1961).

/9/ A. J. F. Boyle, D. St. P. Bunbury, C. E. Edwards, Proc. Phys. Soc., 79, 416 (1962).

/10/ M. F. Crawford and A. L. Schawlow, Phys. Rev., 76, 1310 (1949).

/11/ D. A. Shirley, Phys. Rev., 124, 354 (1961).

/12/ V. A. Bryukhanov, N. N. Delyagin, A. A. Opalenko, V. S. Shpinel', ZhÉTF, 43, 432 (1962).

/13/ Bykov, G. and Pham Zuy Hien, ZhÉTF, 43, 909 (1962).

/14/ A. J. F. Boyle, D. St. P. Bunbury, C. E. Edwards, Proc. Phys. Soc., 77, 1062 (1961).

/15/ N. E. Alekseevskii, Pham Zuy Hien, V. G. Shapiro, V. S. Shpinel', ZhÉTF, 43, 790 (1962).

/16/ H. M. Foley, R. M. Sternheimer, and Jycko, Phys. Rev., 93, 734, (1954); 102, 731 (1956); 84, 244 (1951).

/17/ G. Burns, Phys. Rev., 124, 524 (1961).

/18/ C. Alff and G. K. Wertheim, Phys. Rev., 122, 1414 (1961).

/19/ G. Burns and E. G. Wikner, Phys. Rev., 121, 155 (1961).

/20/ C. H. Tounes and Dailey, J. Chem. Phys., 17, 782 (1949).

/21/ V. A. Bryukhanov, V. I. Gol'danskii, N. N. Delyagin, L. A. Korytko, E. F. Makarov, I. P. Suzdalev, V. S. Shpinel', ZhÉTF, 43, 448 (1962).

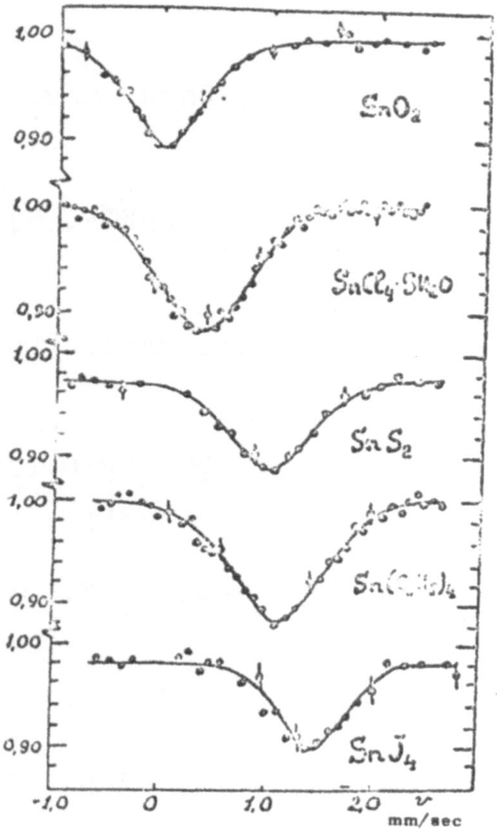

Fig. 1. Resonance absorption spectra for tetra-valent tin compounds obtained with an SnO_2 source. The positions of the maxima for different absorbers are displaced with respect to zero velocity.

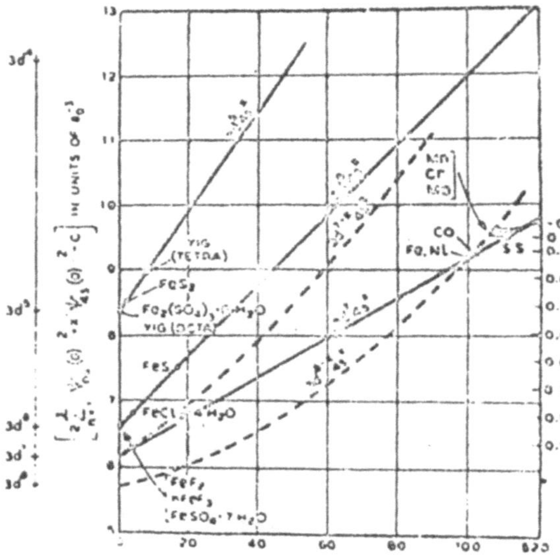

Fig. 2. Possible interpretation of isomeric chemical shifts for different Fe^{57} compounds /7/.

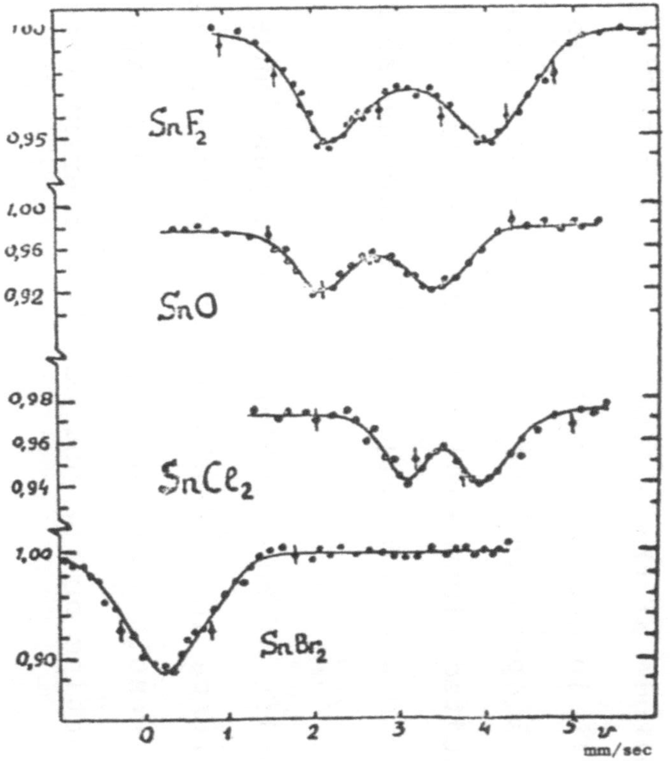

Fig. 3. Resonance absorption spectra for bi-valent tin compounds.

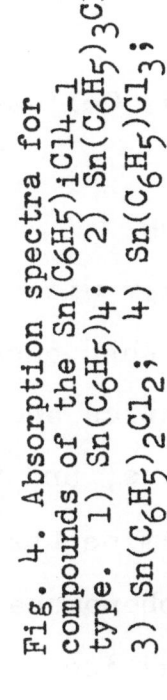

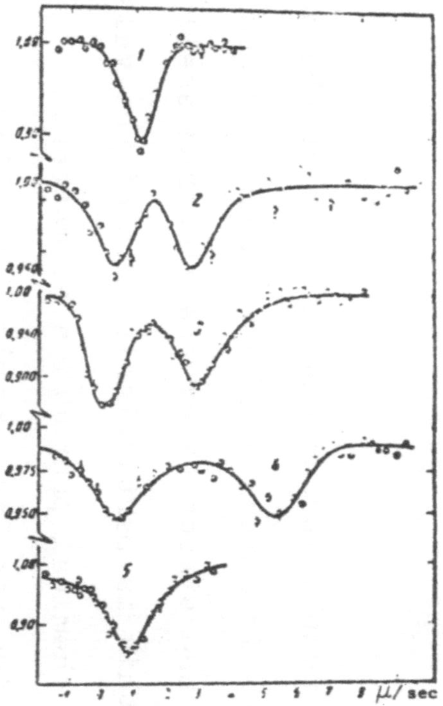

Fig. 4. Absorption spectra for compounds of the $Sn(C_6H_5)_iCl_{4-i}$ type. 1) $Sn(C_6H_5)_4$; 2) $Sn(C_6H_5)_3Cl$; 3) $Sn(C_6H_5)_2Cl_2$; 4) $Sn(C_6H_5)Cl_3$; 5) $SnCl_4$.

APPLICATIONS OF THE MÖSSBAUER EFFECT IN CHEMISTRY

V. I. Gol'danskii

INTRODUCTION

A fairly wide group of phenomena is known at the present time which show that there is a relation between the various nuclear characteristics and the structure of the electron shell surrounding a nucleus, and this is particularly true of the valence shells which determine the chemical properties of matter.

Such phenomena include for example the following:

1) The isotope shift in optical spectra, due to the optical electron interacting with different Coulomb energies with the nuclei of different isotopes of the same element, because they have different dimensions. The magnitude of the shift is $\Delta\lambda/\lambda \approx 10^{-6}$ /1/.

2) Isomer shift in optical spectra, i.e., change in wave length of the optical spectrum line when a transition occurs from the ground to an excited (isomeric) state of the same isotope. This effect comes from the same causes as the isotope shift but is several times weaker /2/.

3) Chemical shift of the nuclear magnetic resonance (NMR) lines, due to difference in screening of the nucleus by valence electrons, with the result that there is a difference in the effective magnetic field acting on the nucleus in different chemical compounds /3/.

4) The relation between the form of the hyperfine structure of the rotational lines or of the nuclear quadrupole resonance

(NQR) spectra and the chemical bonding of the isotopes in question /4/.

The hyperfine splitting of the rotational levels observed in microwave spectroscopy experiments is due to interaction between the nuclear spin and the axis of rotation of the molecule through the nuclear quadrupole moment (Q) and the electric field gradient ($q = \partial^2 V/\partial Z^2$) in the region where the nucleus is located. The transition energy in NQR (radiofrequency region as in NMR) is also determined by the values of Q and q, namely, by the nuclear quadrupole interaction constant $W = eQq$. However, the electric field gradient, q, which occurs in this constant and is a measure of the departure from spherical symmetry of the charge distribution around the nucleus, depends both on the electron shell structure of the particular atom itself, and of the whole molecule, as well as on the crystalline macrostructure.

None of the phenomena recounted above have to do with nuclear transformations. However, in the last ten or fifteen years, studies have also been made on some of the chemical changes in various nuclear transformations (i.e., those due to the electron shell structure of radioactive isotopes).

We can mention in this connection:

1) Chemical change in the lifetime of radioactive isotopes which transform by an electron capture mechanism /5/. The thing is that the electron capture rate constant is proportional to the electron density, $|\Psi(0)|^2$, in the region where the nucleus is located, and this density is essentially dependent on the electron shell structure. The changes of this sort observed are very weak ($\Delta\tau/\tau \lesssim 2 \cdot 10^{-3}$), but they may play a large role under specific condi-

tions, for example, in K-capture in Be^7 nuclei when the beryllium atoms are strongly ionized in a thermonuclear fuel mixture /6/.

2) Chemical changes in the rate of isomeric nuclear transitions, accompanied by strong internal electron conversion /7/ (changes have been observed up to $\Delta\tau/\tau = 2$--$3\cdot10^{-3}$). This effect not only depends on the s-electrons, but also on the state of the many shells with different n and l, and for this reason is very difficult to interpret quantitatively. New and abundant possibilities for observing the effect of chemical structure on nuclear transformations, and hence of getting additional information on chemical structure and the change in chemical properties produced by various factors, have been opened up by the Mössbauer effect /8/. We shall try, in the present paper, to give a general description of the prospects when using the effect for chemical purposes, and outline some of the information which it has already yielded that is of interest to chemists (without however pretending to make a complete literature survey), and, finally, we shall sketch various possible lines of further study.

II. SHIFT AND SPLITTING OF MÖSSBAUER SPECTRUM LINES

The analog of the isotope, or, better, optical spectrum isomer shift in Mössbauer spectra is the shift, which is also usually called isomeric, but which is more correctly called chemical in the present case.

The minimum total energy of a system made up of a nucleus and electron shells occurs for the case of a point nucleus. The farther the electron charge distribution of the nucleus extends, the higher-- for a given electron shell structure-- is the total energy of the system. Accordingly, an increase in the dimensions of an isomeric excited state of a nucleus over the dimensions in the ground state produces an increase in the gamma transition energy. On the other hand, the greater the electron density $|\Psi(0)|^2$ at the nucleus, due principally to s-electrons (the contribution of $P_{\frac{1}{2}}$ electrons is considerably less), the greater also-- for given nuclear dimensions-- will be the total energy of the system. Accordingly, changing $|\Psi(0)|^2$ in the atoms of the absorber from what it is in the emitter also causes them to have different gamma transitions energies.

The change in transition energy resulting from a change in nuclear dimensions between the ground and the excited state is given by the relation /9/:

$$\Delta E = \pi e^2 \frac{a_0^2}{Z} |\Psi(0)|^2 \cdot R^{2\rho} \frac{\Delta R}{R} \left(\frac{2Z}{a_0}\right)^{2\rho} \cdot \frac{1}{\Gamma(2\rho)^2} \cdot \frac{3-2\rho}{3+2\rho}, \qquad (1)$$

where a_0 is the Bohr radius, Z is the nuclear charge, $\rho = \sqrt{1-(Ze^2/\hbar c)^2}$, R is the mean radius of the nuclear electric charge distribution, and $\Delta R = R_e - R_0$ is the change in this radius on going from the ground (R_0) to the excited (R_e) state.

The experimentally measured chemical shift, $\delta = \Delta E_{abs} - \Delta E_{em}$ is proportional to:

$$\delta \backsim (R_e - R_0)\left\{|\Psi(0)|^2_{\text{погл.}} - |\Psi(0)|^2_{\text{изл.}}\right\}, \qquad (2)$$

where a positive shift corresponds with moving the absorber toward the emitter in the spectrum measurement experiments. (We shall not touch here on the additional effect /10/, coming from second order

Doppler shift resulting from having a difference between the Debye temperatures of the absorber and emitter and their temperatures under the conditions of the experiment).

Using Eq. (2) makes it possible first of all to find the sign of the quantity ΔR by comparing the chemical shifts of several absorbers for which it can be said from general chemical considerations which molecules have higher and which have lower electron density at the nucleus. Thus, for example, the positive chemical shift of Fe_2O_3 with respect to metallic iron, if we consider the fact that oxidizing Fe_2O_3 removes the external (4s) electrons , i.e., reduces $|\Psi(0)|^2$, leads to the conclusion that ΔR is negative (i.e., $R_e < R_0$) for Fe^{57} /11/. In a similar way, the positive chemical shift of bivalent tin compounds with respect to quadrivalent tin compounds, in which $|\Psi(0)|^2$ is greater, led to the conclusion in /12/ that ΔR is positive (i.e., $R_e > R_0$) for Sn^{119}.

Even greater possibilities are opened up by Eq. (2) in conjunction with calculating the absolute values of $|\Psi(0)|^2$. Thus, the calculations of $|\Psi(0)|^2$ for states of iron with a different number of 3d-electrons, made by Watson and Freeman /13/ by the Hartree--Fock method, formed the basis for determining the absolute value of $\Delta R/R$ for Fe^{57}, and the subsequent calculations of the contribution made by 4s-electrons to the valence bonds of iron in its various compounds /14/. Later on we shall give a more detailed discussion of both these calculations and similar considerations related to our observations of the Mössbauer effect in tin compounds /15/.

A second source of information on the electron shell structure of Mössbauer atoms is provided by the quadrupole splitting of the spectrum lines.

The interaction between a nucleus with the quadrupole moment Q and an axially symmetric inhomogeneous electric field with gradient q splits the level with moment I into sublevels with different magnetic quantum numbers m, separated by the distance

$$\Delta = eQq\cdot\frac{3m^2 - I(I+1)}{4I(2I-1)}\,. \tag{3}$$

In the most common case of Mössbauer M1 gamma-transitions between $I = \frac{1}{2}$ and $I = 3/2$, splitting the level with $I = 3/2$ into sublevels with $m = \pm 3/2$ and $m = \pm \frac{1}{2}$ produces a spectral doublet, in which one of the lines corresponds with the σ- transition ($\pm\frac{1}{2} \rightarrow \pm \frac{1}{2}$), while the other line corresponds to the π- transition ($\pm\frac{1}{2} \rightarrow \pm 3/2$), and the difference between the energies of the π and σ transitions is $\Delta = \frac{1}{2}eQq = \frac{1}{2}W$. Here, the energy of the π-transition is greater than the energy of the σ-transition if W is positive, and less if W is negative. The magnitude of the quadrupole interaction constant W is itself only a relative measure of the inhomogeneity of the electric field at the Mössbauer nucleus. In order to compare this constant with the values of the field gradient with different assumed concrete electron shell structures, we have to make sometimes quite complicated calculations of the intramolecular electric fields (for example in the sense of Townes and Dailey's calculations /16/), and find the nuclear quadrupole moment experimentally or also by calculation (for a nucleus-- as an example, for Sn^{119} and Fe^{57}-- in an excited state). So far, the analysis of the data on the quadrupole splitting of Mössbauer spectra has only been of a semiquantitative nature. However, a number of interesting results have been obtained, which we shall talk about later on.

The Zeeman splitting picture in Mössbauer spectra is often quite complicated, in view of the fact that in the general case it is determined by two values of the nuclear magnetic moment-- in the ground and in the excited state, and the two (different) values of the local magnetic fields for the nuclei of the emitter and absorber. In addition, the various Zeeman components with different m are shifted differently, by the quadrupole interaction. Nevertheless, interpreting the Zeeman splitting is much simpler, and, if the necessary set of experiments has been performed (without special quantum mechanical calculations), reduces to a direct determination of the absolute values of the magnetic field in the region where the Mössbauer nuclei are located.

III. PRELIMINARY REMARKS ON APPLICATIONS OF THE MÖSSBAUER EFFECT TO CHEMISTRY

a) Frequency of Occurrence of Mössbauer Transitions.

The first question of importance in evaluating the prospects for making chemical applications of the Mössbauer effect is that of the frequency of occurrence of Mössbauer nuclei. As we know, the probability of observing the effect rapidly (in the Debye approximation, exponentially) decreases with increase in the recoil energy $R = E^2/2Mc^2$, which is proportional to the gamma-transition energy E, and inversely proportional to the mass of the emitter. In addition, as a result of the general reduction in nuclear level density (for a given excitation energy) with decrease in nuclear mass, high energy (of the order of MeV) gamma-transitions predominate for light nuclei. Accordingly, we must exclude from the number of even potential Mössbauer emitters and exciters such chemically interesting light element nuclei as carbon, nitrogen, oxygen (not to mention of course hydrogen). The Mössbauer transition energies lie in a range from several keV (with no lower limit in principle) up to several hundred keV. The duration of the corresponding transitions will be neither too small-- where the ratio of the level width Γ to the transition energy E becomes so large that the resonance loses its selectivity-- nor too large, for very small values of Γ/E give low probability and poor observability of recoilless gamma scattering and absorption. The most promising range of values of the ratio Γ/E is 10^{-10}--10^{-14} , although observations of the Mössbauer effect have already been made even for $\Gamma/E = 5.1 \cdot 10^{-16}$ (Zn^{67}) /17/. Lists of possible Mössbauer nuclei are given in a number of reviews /18,19/. Although the group is comparatively restricted (different isotopes of 35 elements are mentioned, from iron to mercury), quite ample possibilities are opened up for various chemical studies, with particular interest-- from the point of view of the chemist-- being presented by the Mössbauer effect in the study of complex and metalloorganic compounds. We shall now give some of the elements for which recoilless resonance gamma-fluorescence has been observed (underlined) or should be observed: <u>iron</u>, <u>nickel</u>, <u>zinc</u>, germanium, arsenic, ruthenium, <u>tin</u>, antimony, <u>tellurium</u>, <u>iodine</u>, cesium, almost all the lanthanides (the effect has already been observed for Sm, Dy, Er, Tm), hafnium, tantalum, tungsten, <u>rhenium</u>, osmium, <u>iridium</u>, platinum, <u>gold</u>, and mercury. To this list can undoubtedly be added many of the transuranium elements. To the advantages of the Mössbauer effect over the NQR method must also be added the possibility of using quadrupole moments not only for ground states, but for excited nuclear states as well. For example, the nuclei of all the stable isotopes of iron, tin, tellurium, and tungsten, have no quadrupole moment at all in the ground state, so that the NQR method is completely inapplicable to these elements. However, in recoilless resonance fluorescence, it is possible to have quadrupole splitting of the excited levels with $I \gg 1$ (or $I \gg 3/2$) in all these cases.

b) Role Played by a Light Atom Environment.

The next important question is whether or not it is possible to observe the Mössbauer effect in compounds in which the Mössbauer atoms are embedded in molecules or lattices constructed mainly of much lighter atoms, such as hydrogen, carbon, nitrogen, oxygen, etc. In view of the fact that recoil of the Mössbauer nucleus produces combined oscillation of a large number of atoms, it would seem that the probability of observing recoilless gamma resonances in complex systems would be determined (in the Debye approximation) not by the masses of the emitters and absorbers themselves but by those of the basic, lighter components. If the light atoms were not only incapable of being Mössbauer themselves, but in addition annihilated the effect by their presence, the chances of using recoilless resonance fluorescence in chemistry would be poor indeed. However, it has also been pointed out in the review by F. L. Shapiro /20/ that finding the probability of the Mössbauer effect from the value of the mass averaged over the masses of all the atoms in the lattice is based on neglecting the role played by the high frequency part of the optical branches of the vibration spectrum, ~~although the reason~~ the justification for ~~this~~ which is far from obvious.

A consistent theory of the role played by the optical branches of the phonon spectra in the Mössbauer effect has been worked out by Yu. M. Kagan /21,22/, who showed that it is not right to use the Debye temperature as even an approximate characteristic of recoilless gamma fluorescence. Yu. M. Kagan has also shown that as a result of the large amplitudes of the optical branches, there will be both a large probability of the Mössbauer effect in light systems with heavy emitters or absorbers, and an anomalous slow temperature dependence of the effect. Actually, as was first shown in the experiments of Wertheim /23/ and Ruby /24/ there is even a considerable probability of observing the Mössbauer effect in Fe^{57} in ferricyanides, for example in $Na_4Fe(CN)_6 \cdot 10H_2O$, where the mean mass of the atoms is a factor of 5.5 less than that of the atoms absorbing the γ- quanta. However, these experiments, which were only made at one, and at that quite low temperature (78° K), have still not made it possible to tell anything about the general properties of the effect when the Mössbauer nuclei are in a light environment. Accordingly, much more interest attaches to the results of the experiments mentioned in Yu. M. Kagan's first paper /21/, made by V. V. Sklyarevskii, B. N. Samoilov and E. P. Stepanov on Dy_2O_3 /25/, and by V. S. Shpinel' and co-workers on SnO_2 /26/, who observed both a large probability of the Mössbauer effect, and an anomalous weak temperature dependence in these oxides. Particularly instructive in this respect is the result of our work /27/, in which it was shown that there is a quite large effect in the polymer containing tin (based on methylmethacrylate) of the composition:

$$(C_2H_5)_3 \ Sn - O - C \begin{smallmatrix} \nearrow O \\ \searrow \\ C = CH_2 \\ | \\ CH_3 \end{smallmatrix} \ + \ H_3C - O - C \begin{smallmatrix} \nearrow O \\ \searrow \\ C = CH_2 \\ | \\ CH_3 \end{smallmatrix}$$

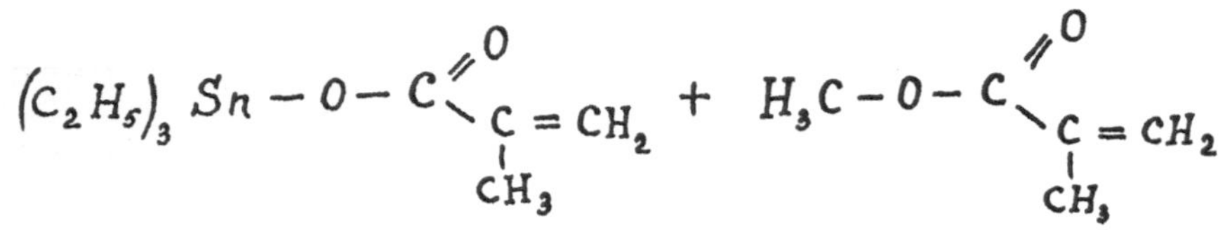

the mean atomic weight of which is a factor of 15 less than that of the tin. The Mössbauer spectrum of this polymer at $t^0 = 79^0$ K and 195^0 K is shown in Fig. 1. The recoilless resonance absorption probability (f') was reduced by not more than a factor of two in this range of temperatures. The theoretical and experimental data just given are sufficient to rid one of any skepticism with regard to the possibility of applying the Mössbauer effect to a very large group of chemical compounds.

c) Delimitation of the Roles Played by the Nearest Chemical Bonds and by the Macrostructure.

One more question that is important to the outlook for chemical applications is that it is possible to separate the features in the Mössbauer spectra that come from the structure of the various molecules under study from those which come from the macroscopic properties of the lattice, (which, it is true, themselves depend to some extent on molecular structure). It has been demonstrated beyond doubt a number of times that in the general case some role is being played by the macrostructure, and that there is a definite effect of intermolecular and collective interactions, i.e., of the charge fields and magnetic moments occurring in the immediate vitcinity of the Mössbauer atoms. It is sufficient to recall for example the differences in quadrupole splitting between tetragonal (W $= 2.5 \cdot 10^{-7}$ eV) and rhombic (W $= 3.5 \cdot 10^{-7}$ eV) tin oxide, SnO /28/, or the differences (of $4.7 \cdot 10^{-8}$ eV) between the chemical shifts in iron and gray tin /28/. The macrostructure effect shows up in the temperature changes in the magnitudes of both the chemical shift and the quadrupole splitting. It is however also immediately obvious that "inert" systems may be found, in which the macrostructure does not prevent making a study of the properties of the individual molecules, which are of most interest to chemists.

As an illustration of these possibilities, we can cite some results which we have obtained from observing the Mössbauer effect in tin compounds /29/. Figure 2 shows the line found in /29/ for a glass containing 9.1% SnO_2 along with silicon, boron, sodium, and aluminum oxides, the radiator being crystalline tin dioxide, SnO_2. In addition to the fact that there is a strong effect (of 77° K) in the absorber, which until recently was assumed to be typically amorphous, it is interesting to see that there is no chemical shift, in spite of the transition from pure SnO_2 to solid solution. This is all the more remarkable for the reason that it is natural to assume that the solid and involatile oxide SnO_2 does not exist in the form of separate molecules (like CO_2), but rather as a single macromolecule-- a sort of umlimited polymer, in which each tin atom is bound with four oxygen atoms, and each oxygen atom is bound with two tin atoms. This means that when the glass was made, the continuous - Sn--O bonds were not destroyed, and the unlimited polymer structure was maintained. Other examples of inert matrices, in which no distortion of the spectral characteristics is observed, are provided by the study made in /29/ on solutions of molecular crystals of organic tin compounds in inert organic solvents.

Thus, no changes are noted in either the chemical shift or the quadrupole splitting when going from crystalline $Sn(C_6H_5)_4$ to a solution in polymethylmethacrylate polymer, or going from $(C_2H_5)SnCl_2$ to a solution in dichlorethane. Accordingly, in all these cases the decisive effect on the form of the spectra is exerted by the chemical bonds nearest the tin atom. This is of two-fold interest to chemists, since it is possible to investigate molecular properties in pure form (in inert solvents), and at the same time possible to study the effect of the solvents themselves, as a function of polarity, their ability to enter into chemical reaction with the dissolved substance, etc. That it is possible to detect the effect of even the nearest chemical bonds in the Mössbauer spectra (not to mention eliminating the macrostructure) is shown clearly by the spectra given in Fig. 3, which are practically identical for the tin polymer mentioned above /27/ and for triethyltin acetic ester $(C_2H_5)_3Sn\text{--}O\text{--}C\begin{smallmatrix}\nearrow O \\ \searrow CH_3\end{smallmatrix}$, in which the bonds nearest the tin (right up to the methyl group) are the same as in the polymer. The spectra were also the same for tetraphenyl tin, $Sn(C_6H_5)_4$, and triphenylstyryl tin $(C_6H_5)_3Sn\text{--}C_6H_4\text{--}CH{=}CH_2$, where there is a vinyl group in the para position in one of the benzene rings, but all four of the valence bonds of the tin itself are the same.

Thus, under definite conditions, the Mössbauer spectra are characteristic of precisely the nearest chemical bonds of the emitting or absorbing atoms. This opens up the possibility of investigating the nature and mutual effect of these bonds (which we shall speak of later), and at the same time provides ways of discovering and making quantitative calculations on the effect that more distant bonds have on the electron shell structure of the Mössbauer atoms.

d) Possible Complications from a Preceding Beta Decay or Gamma Emission.

Since the Mössbauer emitting nucleus is formed as a product of a preceding beta decay (for example $Co^{57} \xrightarrow{\beta^-} Fe^{57}$) or high energy gamma-emission (for example $Sn^{119m} \xrightarrow[\gamma]{65keV} Sn^{119*}$), the chemical state of the emitter atoms is, in the general case, by no means unique. As a matter of fact, the "stirring up" of the electron shells that occurs during beta decay and gamma-transitions produces multiple ionization of the emitting atoms, and this causes them to be in the most diverse--including metastable--chemical states. Theoretical calculations of the relaxation time of these metastable states give very small values as compared with the duration of the subsequent gamma-transitions ($\sim 10^{-12}$ sec) /30/. However, various chemically stable states can also be formed in the emitter atoms, with different chemical shifts in the Mössbauer gamma-transition energy, and as a result the form of the absorption spectra will be distorted. It was shown in a recent paper by Wertheim /30/ that when using a Co^{57} source in the form of cobalt oxide, CoO, that the instant the following gamma-emission occurs, the Fe^{57} atoms are in both the $Fe^{\cdot\cdot}$ state (forming an oxide isomorphic with CoO) and the $Fe^{\cdot\cdot\cdot}$ state, which produces appreciable splitting of the Mössbauer line (see Fig. 4.), as well as complication in the Zeeman effect picture.

Thus, on the one hand, there is the problem of selecting an emitting material that will give the purest "single state" Mössbauer atoms (of especial importance in chemical investigations)-- with iron, for example, obviously stainless steel, while on the other hand the Mössbauer effect presents obvious possibilities for making direct observations of the various chemical consequences of nuclear transformations, such as the formation, maintenance, and healing of all possible kinds of radiation defects.

e) Asymmetric Doublet Splitting in Mössbauer Spectra.

It may be seen from Fig. 4 that having two chemical forms present results in asymmetric splitting of the Mössbauer spectrum. It is perfectly natural to have asymmetry of this sort in all cases where the doublet comes from two different chemical states of the Mössbauer atoms, i.e., the doublet is formed by two singlet lines with different chemical shifts. Examples of such structures, where iron is performing two functions, observed by means of the Mössbauer effect /14,40,41/, are provided by certain mixed oxides (for example, titanium-iron or yttrium-iron garnets), where the iron is nominally trivalent, but has a coordination number of 4 or 6 (tetrahedral and octahedral lattices). It is true, that in these cases the form of the spectra is affected by the Zeeman and quadrupole splitting in addition to the double value of the chemical shifts. However, having asymmetric doublet splitting in the Mössbauer spectra is in itself no proof that the emitting or absorbing atoms are performing a double chemical function (such as we had assumed, for example, at the start to be the case in the organic tin compounds of /29/, or as was done in a recent paper /44/ for hemin for the same reason). It has been shown in /15,34,35/ that asymmetric quadrupole splitting of the Mössbauer spectra will be observed even in perfectly isotropic polycrystalline samples, if there are polycrystals in which the Mössbauer effect is anisotropic /36,37/. Thus, to prove that the asymmetric splitting is due to two singlet lines with different chemical shifts, and is not quadrupole, we have to be sure that the splitting is independent of the orientation of the sample (or of the angle at which an anisotropic sample is located) with respect to the direction of the gamma-quanta.

IV. PRINCIPAL RESULTS OF STUDIES OF THE MÖSSBAUER EFFECT IN IRON AND TIN COMPOUNDS.

The most complete Mössbauer effect studies made so far are those on iron and tin compounds using Co^{57} and Sn^{119m} sources. Before discussing the principal results obtained on iron compounds, we shall give a compilation of the data in Figs. 5 and 6. Since the literature references are only shown on the figures when different results were obtained for the same or similar compounds in different papers, we shall give the principal literature sources here. The spectra of iron in different metals and alloys were taken in /14,23,60,61/. Various iron salts were studied, in addition to /14/, in the papers by de Benedetti et.al. /38/ and by Kerler and Neuwirth /39/. In /14,40,41/ studies were made of simple and mixed (for example, garnet) iron oxides. The nitride Fe_4N formed the subject of /42/, and the two forms of the sulfide FeS_2-- pyrite and marcasite-- were dealt with in the paper by Solomon /45/. The spectra of rare earth ferrides were observed by Wertheim and Wernick in /31/, UFe_2 in a paper by a group of Japanese authors /32/, and other intermetallic iron compounds in /33/. The papers by Zahn, Kienle, and Eicher /43/, and by Epstein /44/ dealt with the Mössbauer spectra of ferrocene and of the ferricenium cation, while in /44/ a study was also made of cyanide iron complexes (see in this connection also /14/24/, as well as of carbonyl, phthalocyanine, hemin, acetylacetonate, and other complex iron compounds. The complete data on chemical shifts (relative to Fe^{57} in stainless steel) are given in Fig. 5, while the information on quadrupole splitting (the distances between the two doublet peaks) is given in Fig. 6. The data are given for room temperature, but where a study was made of the temperature dependence of δ or Δ (in /38,39,31/)vertical lines are drawn on the figures giving the upper and lower limits of the temperature range.

It may be seen from Fig. 6 that the quadrupole splitting sometimes increases by quite a bit when the temperature is lowered. As for the temperature dependence of the chemical shifts, it can be practically wholly explained not by the change in $(|\Psi(0)|^2)$, (i.e., by the actual chemical shift), but by the second order Doppler effect /10/ associated with the characteristic lattice vibration energies. Only in special cases (for example ferricyanides) is it possible to imagine that there will be some change in $(|\Psi_{1,2,3s}(0)|^2)$ as a result of temperature changes in the 3d-electron screening effect /39/.

It is obvious from Fig. 5 that according to the magnitude of the chemical shift, inorganic iron compounds divide into three groups, corresponding to metallic, bivalent, and trivalent iron. Since the electron density at the place where the nucleus is located is assumed to be greater in metallic iron than in the trivalent state (which corresponds to $\Delta R(Fe^{57}) < 0$ /11/), the value of $|\Psi(0)|^2$ is least for bivalent iron. The treatment of these results in /14/, based on the calculations of Watson and Freeman /13/, leads to the conclusion that the electron configuration in metallic iron (Fe in stainless steel) is $3d^7 4s^x$, while both trivalent and bivalent iron compounds have electron shells of the type $3d^5 4s^x$ and

$3d^6 4s^x$ respectively, where $\underline{x}$ is the fraction of the 4s vacancies in the shells of the iron ions filled by electrons from the anions bound with them. These conclusions are illustrated in Fig. 7, where the left-hand ordinates are the calculated values /13/ of $2\,[\sum_{i=1,2,3} |\Psi_{is}(0)|^2]^- C$, i.e., the total densities of 1,2, and 3s-electrons at the place where the nucleus is located, for a different number ($n = 4,5,6,7,8$) of 3d electrons. The straight lines $3d^n 4s^x$ for $n = 5,6,7$ give the additions to $2 \sum_{i=1,2,3} |\Psi_{is}(0)|^2$ with filled $1s^2 2s^2 p^6 3s^2 p^6 d^n$ configurations from 4s-electrons. The calculation is made on the assumption that these additions are simply $X|\Psi_{4s}(0)|^2$, where the magnitude of $|\Psi_{4s}(0)|^2$ is found by the Fermi--Segrè--Goudsmit method /46,47/ (here no account is taken of 4s-electron screening of the internal s-shells). In choosing the scale of the chemical shifts (and to define the value of $\Delta R/R$), it is assumed that the compounds having the largest values of δ in a given valence group, namely $Fe_2(SO_4)_3 \cdot 6H_2O$ and the garnet $3Y_2O_3 \cdot 5Fe_2O_3$ for $Fe^{\cdot\cdot\cdot}$, and FeF_2, $KFeF_3$ and $FeSO_4 \cdot 7H_2O$ for $Fe^{\cdot\cdot}$ are the "standards", having pure $3d^5$ and $3d^6$ electron configurations. These calculations give the value $\Delta R/R = -1.8 \cdot 10^{-3}$, and the values of $\underline{x}$ in the $3d^n 4s^x$ configurations for all the other compounds in the $Fe^{\cdot\cdot\cdot}$ and $Fe^{\cdot\cdot}$ series. The data for ferro- and ferricyanides in /14/ is not being analyzed.

It must however be noted that ascribing the $3d^5({}^6S_{5/2})$ configuration to the "standard" $Fe^{\cdot\cdot\cdot}$ and $3d^6({}^5D_4)$ to $Fe^{\cdot\cdot}$ is not in accordance with the data on the fine structure of the edges of the X-ray K-absorption spectra, which data make it possible to find the effective charges η of the various atoms in chemical compounds /48,49,50/. In order to draw any conclusions on the electron shell structure, we must combine the results of effective charge experiments, giving the total number of valence electrons, with the observations of Mössbauer spectra, which make it possible to differentiate between the contributions from s-(chemical shifts), p- and d-(quadrupole splitting) electrons. In the case of iron, $\eta = +1.9$ for $Fe^{\cdot\cdot}$ salts, +1.0 for $K_3Fe(CN)_6$, and +0.4 for the carbonyl, $Fe(CO)_5$, and ferrocene, $Fe(C_5H_5)_2$, the values of η for ferrocene and the ferricenium cation being practically the same /50/. From the analogy between iron and cobalt, we should take $\eta = +1.2$ for $Fe^{\cdot\cdot\cdot}$, i.e., the effective charge for trivalent iron is not greater, but less, than for bivalent iron.

Assuming now that the total number of valence electrons is $A = 8 - \eta$, and that $A = 8$ for metallic iron, we can interpret the chemical shift data for iron (in the light of Watson and Freeman's calculations /13/) in the same way as was done in Fig. 8. Then we must assume a structure for metallic iron, which is close to $3d^6 4s^2$ (and not $3d^7 4s$), for $Fe^{\cdot\cdot\cdot}$--$3d^{6,8-x}4s^x$, where $x = 1$--1.25, and for $Fe^{\cdot\cdot}$--$3d^{6,1-x}4s^x$, where $x = 0$--0.35. In the same way, for Fe^{57}, we obtain $\Delta R/R = -5 \cdot 10^{-4}$ (instead of $-1.8 \cdot 10^{-3}$ /14/). Figure 8 also gives points for the iron complex compounds, corresponding to the configurations $3d^{5.5}4s^{1.5}$ for $K_3Fe(CN)_6$, about $3d^{6.2}4s^{1.4}$ for ferrocene and ferricenium, and $3d^{5.8}4s^{1.8}$ for carbonyl. However, this would be too rough an interpretation, since it does not include sp^3d hybridization, which shows up when iron coordination bonds are formed in complex compounds. The simple fact of part of the 3d-electrons going to 4p-orbits would, as may be seen from Fig. 7, increase $|\Psi_{1,2,3,5}(0)|^2$, i.e., it would reduce the positive chemical shifts, as is observed in all complex compounds. It is now easily seen that having assumed a certain number of electrons to pass to the 4p-orbit, we cannot reconcile this assumption with the observed values of the chemical shifts from 3d-electrons alone, and it is absolutely necessary to reduce the number of both 3d and 4s-electrons at the same time (the curves in Fig. 8 shift to the left and up). The possibility of having the electrons arranged in different ways in the 3d and 4sp-orbits enables us, in principle, to give a simple explanation of why the values of δ are very nearly the same for different complexes where the valence of the iron is nominally different. For the explanations to be anywhere near right, we must, however, keep in mind both the chemical shifts and the effective charges, as well as the quadrupole splitting of the Mössbauer spectra. It may be seen from Fig. 6 that the splitting is considerably greater for $Fe^{\cdot\cdot}$ than for $Fe^{\cdot\cdot\cdot}$ salts. At first glance, this may be accounted for by the spherical symmetry of the fundamental $Fe^{\cdot\cdot}$ term $(6S_{5/2})$ in contrast with $Fe^{\cdot\cdot}$ $(^5D_4)$. However, this explanation assumes an effective charge of +3 for $Fe^{\cdot\cdot\cdot}$ and +2 for $Fe^{\cdot\cdot}$, and is thus in contradiction with the information given above on the effective charge of the iron. Further, the temperature dependence of the quadrupole splitting is greater for $Fe^{\cdot\cdot}$ than for $Fe^{\cdot\cdot\cdot}$ salts, which is accounted for in /38/ by thermal excitation of the higher lying $Fe^{\cdot\cdot}$ terms. However, in our opinion, strong temperature dependence resulting from various phase transitions, for example, those associated with change in crystal hydration states, is more to be expected for quadrupole splitting due, not to the structure of the actual electron shells, but to inhomogeneous macrostructure fields. Accordingly, the dependence will have more effect on an ion with spherically symmetric electron shells, where all the splitting is due to the effect of the remote environment. The specific effect of the structure of iron cyanide complexes, which makes them different from ordinary $Fe^{\cdot\cdot\cdot}$ and $Fe^{\cdot\cdot}$ salts, also shows up in the values of the quadrupole splitting shown in Fig. 6. Red potassium ferricyanide shows rather small quadrupole splitting (like the usual $Fe^{\cdot\cdot\cdot}$ salts), and yellow potassium ferrocyanide shows no splitting at all ($\Delta < 0.1$ mm/sec), while for ordinary $Fe^{\cdot\cdot}$ salts (and sodium nitroprusside) the distance between the two peaks is greater than 1.5 mm/sec. Nearly the same as the cyanide complexes of iron both in the magnitude of the chemical shift and in the quadrupole splitting of the Mössbauer spectra is /44/ hemin (ferrihem) a complex of oxidized (trivalent) iron with protoporphyrin. At the same time, a structure very similar to porphyrins--phthalocyanine $Fe^{\cdot\cdot}$--shows values of chemical shift and quadrupole splitting considerably greater than for hemin. It is interesting, finally, to note that the dipyridine--acetylacetonate of bivalent iron $[Fe^{\cdot\cdot}(AcAc)_2(C_5H_5N)_2]$ in both the value of δ and in the quadrupole splitting falls in the range of trivalent iron salts /44/. Another way of accounting for the quadrupole splitting in some iron cyanide complexes is given in /39/, where, of the five 3d-orbits, two are used for sp^3d^2-hybridization in the bonds formed with the electrons of the bound radicals, while the remaining three are entirely filled by the six electrons of $Fe^{\cdot\cdot}$ (in the ferricyanide $[Fe(CN)_6]^{''''}$). The result is to produce a spherically symmetric charge distribution, with no quadrupole splitting. However, the idea of having no quadrupole splitting when the

3d-shell is completely filled (i.e., the contribution from the σ, π, and δ-bands of this shell is cancelled out) is in contradiction with the facts, in that there is strong splitting of the Mössbauer spectra of iron carbonyl and ferrocene.

The presence of weak splitting for $K_3Fe(CN)_6$ (monoclinic crystals) is accounted for in /39/ by the formation of a "hole" in the 3d-shell on going from $Fe^{\cdot\cdot}$ to $Fe^{\cdot\cdot\cdot}$. As we shall soon see, the formation of such a "hole" in the ionization of ferrocene leads, as a matter of fact, to a diametrically opposite result, -- the splitting does not increase, but decreases. Finally, the relatively strong splitting in nitroprusside is explained in /39/ by the replacement of the CN^- radical with NO^+ (with bivalent iron). It is interesting to note that literally the same experimental data obtained for nitroprusside in /33/ is there treated in a completely different way-- as a replacement of CN^- by NO (with quadrivalent iron).

We turn now to ferrocene. The most characteristic thing here is the great difference between the quadrupole level splitting in ferrocene itself, where the splitting is very strong (Δ = 2.33--2.4 mm/sec /43,44/), and ferrocene compounds, where the splitting is much weaker, and, moreover, apparently depends on the nature of the anion (Δ = 0 for the picrate Fn^+ /44/, and Δ = 0.75 mm/sec for Fn^+BF_4'/43/). However, the chemical shifts lie in the range of values characteristic of $Fe^{\cdot\cdot\cdot}$, and are nearly the same for ferrocene (0.46 /44/--0.52 /49/ mm/sec) and Fn^+-picrate (0.53 mm/sec), and are somewhat less (0.25 mm/sec) for Fn^+BF_4'.

An interpretation of the quadrupole splitting data has been advanced by the authors of /43/. As we know, the $Fe(C_5H_5)_2$ contains 18 valence electrons (8 from iron and 5 from each ring), which corresponds to completely filled iron shells. Here, there would be no quadrupole splitting at all, if the σ, π, and δ-bands in the group of d-electrons (i.e., the group of electrons with magnetic quantum numbers $m_l = 0, \pm 1$, and ± 2) had the same values, $\overline{r^3}$, of the mean cube of the distances from the iron nucleus, since the contribution of one d-electron of each band to the value of Δ is proportional to $(m_l^2 - 2)/\overline{r^3_{ml}}$, and $2(0^2 - 2) + 4(1^2 - 2) + 4(2^2 - 2) = 0$. A stronger bond with the radicals corresponds to an increase in $\overline{r^3}$, and thus the authors of /43/ account for the observed quadrupole splitting (assuming that $\Delta > 0$, i.e., the π-transition energy is greater than that of a σ-transition), by saying that the δ-electrons of the iron are more weakly represented in the vicinity of the radicals than the σ and π-electrons, and so they have a larger value for the factor $1/\overline{r^3}$. The reduction in the value of Δ on going from ferrocene to the ferricenium salt $[Fe(C_5H_5)_2]^+ BF_4'$ is explained by the authors of /43/ by saying that when ferrocene is ionized, a d-electron is taken out of a δ-band, so that the electron is "closest" to the iron and thus relatively weakly bound with the rings, so as to make a positive contribution to the quadrupole splitting.

In interpreting the data of /43/ on ferrocene and ferricenium, it is convenient to use the results of the only quantitative calculations of their kind on these complexes made by E. M. Shustorovich and M. E. Dyatkina /51,52/ using the self-consistent molecular orbit (MO) method*. The valence shells of the complexes consist

---

*The author is grateful to E. M. Shustorovich for a detailed discussion of this question.

of nine molecular binding orbits of various symmetries, of the form $\varphi = a\Psi + b\chi$ where the Ψ's are the atomic orbits of iron, the χ's are the MO of the bound rings, and a and b are numerical coefficients, with $a^2 + b^2 = 1$. The results of the calculations are given in the table, where values of a are given for each of the MO's of the complexes.

Symmetry of MO of Complex	Ferrocene /51/ Fn: Form of a Ψ	Ferrocene /51/ Fn: No. of electrons in MO	Ferricenium /52/ Fn^+: Form of a Ψ	Ferricenium /52/ Fn^+: No of electrons in MO
A_{1g}	0.49 s	2	0.49 s	2
	1.00 d_{z^2}	2	1.00 d_{z^2}	2
A_{1u}	0.10 p_z	2	0.09 p_z	2
E_{1u}	0.59 p_x	4	0.60 p_x	4
	0.59 p_y		0.60 p_y	
E_{1g}	0.37 d_{xz}	4	0.45 d_{xz}	4
	0.37 d_{yz}		0.45 d_{yz}	
E_{2g}	0.85 $d_{x^2-y^2}$	4	0.94 $d_{x^2-y^2}$	3
	0.85 d_{xy}		0.94 d_{xy}	

It is obvious that the density of the 4s-electron cloud (on account of the A_{1g} orbit) is practically the same in Fn and Fn^+ -- $(2 \cdot 0.49^2 \approx 0.5)$, and this means that the chemical shifts are about the same in both complexes. The reduction in the quadrupole splitting in ferricenium not only comes from the reduction in the E_{2g} contribution $(3 \cdot 0.94^2 - 4 \cdot 0.85^2 = -0.26)$, but also from the increase in the E_{1g} contribution $(4 \cdot 0.45^2 - 4 \cdot 0.37^2 = 0.24)$ in Fn^+, since E_{2g} gives positive $(m_l^2 - 2 = +2)$, and E_{1g} gives negative $(m_l^2 - 2 = -1)$ splitting. It may be seen from the table that the

E_{2g}-electron removed belongs by more than 70% to the atomic orbit of iron, i.e., the factor $\overline{1/r^3}$ for this electron may quite naturally be assumed greater than for the electrons making a negative contribution to the quadrupole splitting. However, the rearrangement occurring when the E_{2g}-electron is removed not only leads (in two ways), as has already been said, to a reduction in quadrupole splitting, but also serves to maintain the effective charge of the iron atom. The distribution of the total number of electrons in the iron atoms in ferrocene /7,3) and in ferricenium /7,4/ is, from Shustorovich--Dyatkina /51,52/, of the form $3d^{5.4}4s^{0.5}\,p^{1.4}$ ($p^{1.5}$ for ferricenium). It is interesting that if we neglect the effect of the 4p-electrons on $|\Psi_{1,2,3s}(0)|^2$ and consider only the $3d^{5.4}4s^{0.5}$ structure, a calculation of $|\Psi(0)|^2$ for this structure by the Watson--Freeman /13/ method (in the sense of Fig. 8) gives beautiful agreement with the chemical shifts observed for ferrocene and ferricenium. This is an additional proof in favor of the correctness of the theoretical calculation of the ferrocene structure given in the papers by E. M. Shustorovich and M. E. Dyatkina /51/52/.

In concluding our brief analysis of the data on iron, let us stop to consider the values of the chemical shifts for $FeCl_3 \cdot 6H_2O$, which, as may be seen from Fig. 5, fall far out of the series of trivalent iron salts. The value given for this salt in /39/ is $\delta = +0.98 \pm 0.03$ mm/sec, and there is also talk of a "secondary" line, $\delta = +0.03 \pm 0.05$ mm/sec. The reason for distinguishing between a main and a secondary line is that the difference in resonance effect between them is more than a factor of one and a half (15.9% in the main, and 10.2% in the secondary line), i.e., asymmetric doublet splitting is observed which the authors of /39/ were not able to interpret as quadrupole splitting. However, in the light of our papers already mentioned /15,34,35/, in which the asymmetric quadrupole splitting for isotropic polycrystalline samples is accounted for by the anisotropy in the Mössbauer effect in the monocrystals themselves, the case of $FeCl_3 \cdot 6H_2O$ finds a natural explanation, --the chemical shift is $\delta = +0.50$ mm/sec, which is the same as the data of /38/ for anhydrous $FeCl_3$, but the value of Δ(0.95 mm/sec) also falls in the range for trivalent iron salts, although at the very edge of the range (the difference between Δ for iron chloride with water of crystallization, and for the anhydrous chloride, where $\Delta = 0$, is still in need of further clarification).

We turn now to the data for tin. A compilation similar to Figs. 5 and 6 of the results on chemical shifts and quadrupole splitting is shown schematically in Fig. 9. The data are taken from our papers /15,27,29/, as well as from other papers by Soviet /12,53/ and foreign /28/ authors. The figure not only contains all the information on the chemical shifts and quadrupole splittings of the spectra of organic tin compounds (Ph means phenyl, C_6H_5, Et means ethyl, C_2H_5, Pr means propyl, C_3H_7, Bu means butyl, C_4H_9), but the data given provide satisfactory information on the position of the range of values of δ and Δ characteristic of organic tin. The literature references are only given in Fig. 9 for the case where the data are in substantial contradiction, as occurs for example for bivalent tin compounds in the data of Boyle /28/ and V. S. Shpinel' et.al. /12,53/. If a tin compound is only shown in the graph of chemical shifts but not in the quadrupole splitting graph (lower), this means that $\Delta = 0$ for the compound.

What are the basic features of the data shown in Fig. 9? As a rule, all the quadrivalent tin compounds fall in the range of negative chemical shifts (with respect to β-Sn), while bivalent tin compounds are in the range $\delta > 0$. With all the valence electrons present in the tin atom, it is natural to assume a completely hybridized structure of the type $5sp^3$. The wave function of each of the four hybridized orbits is of the form $\psi_{hybr} = \frac{1}{2}\psi_s + (\sqrt{3}/2)\psi_{pz}$, i.e., the contribution from each such orbit to the electron density at the region where the nucleus is located is equal to $(1/4)|\psi_{5s}(0)|^2$. Accordingly, neglecting the perturbing effect of the 5p-electrons on $\sum_{i=1\ldots4}|\psi_{is}(0)|^2$, we may conclude that with four covalent bonds, the electron density at zero corresponds to something like having one s-electron present. With completely ionized $Sn^{::}$, this electron is taken away, and what remains is the structure of completely filled n=1,2,3, and 4 spd-shells. Accordingly, the value of $|\psi(0)|^2$ decreases below what it was with entirely covalent bonds. On the other hand, in completely ionized bivalent tin, the hybridization is destroyed, and two 5p-electrons are taken away, leaving two 5s-electrons, i.e., the value of $|\psi(0)|^2$ increases above the value for completely covalent compounds, but also above the value for completely ionic compounds of quadrivalent tin. Using a similar argument, we not only find the sign of $\frac{\Delta R}{R}$ $(Sn^{119*}) > 0$ without difficulty, but also arrive at the conclusion that strengthening the ionic character of the quadrivalent tin bonds corresponds with increasing the chemical shift in the negative direction (from SnI_4 to SnF_4), while for bivalent tin, strengthening the ionic character of the bonds increases the chemical shift in the positive direction--from SnO to $SnCl_2$. An analysis of the data on bivalent tin compounds has so far been rendered difficult by the fact that the data are obviously in contradiction, see, for example, in Fig. 9, how much difference there is between the data of /28/ and /12,53/ on the chemical shifts for SnF_2, $SnCl_2$, and $SnCl_2 \cdot 2H_2O$ (where for the latter compound there are even contradictions in the value of Δ), and how much departure there is from the general law in the case of $SnBr_2$, for which a strong negative shift with respect to β- Sn was found in /53/. When it comes to quadrivalent tin, we can already make a number of qualitative and even semiquantitative deductions.

A calculation of $|\Psi(0)|^2$ for a single 5s-electron in the field of the skeleton consisting of the tin nucleus and the electrons in the completely filled 1,2,3, and 4-spd shells, made by the Fermi--Segre method /46/, gives the value 1.5·10 /26/ $1/cm^3$ /28/. There are two more possible ways of making the calculation. Boyle et.al. /28/ assume that SnF_4 is an example of a completely ionized compound of quadrivalent, and $SnCl_2$ of bivalent tin. Then these two compounds differ by two 5s-electrons, and the chemical shift between them is equal to 5.2 mm/sec $= 4.1 \cdot 10^{-7}$ eV.

According to Eqs. (1) and (2), the chemical shift for Sn^{119} is equal to:

$$\delta_{Sn} = 1.55 \cdot 10^{-29} \frac{\Delta R}{R} \left\{ |\Psi(0)|^2_{abs.} - |\Psi(0)|^2_{emit.} \right\} \text{ eV} \qquad (4)$$

In /28/, a value is given of $\Delta R/R = 1.1 \cdot 10^{-4}$, which corresponds with a difference in $|\Psi(0)|^2$ between $SnCl_2$ and SnF_4 of $2.4 \cdot 10^{26}$

$1/cm^3$, although this value is not given directly in /28/ but, judging from the text, it was found from including mutual screening of the two 5s-electrons, and is thus less than $2|\psi_{5s}(0)|^2 = 3 \cdot 10^{26}\ 1/cm^3$. The weak point in this calculation is assuming complete ionization in SnF_4 and $SnCl_2$. Accordingly, in our work /15/, we first of all compared the value of the chemical shifts in quadrivalent tin compounds of the type SnX_4 with the electronegativity of the X atoms and with the degree of ionicity of the bond found for Sn--Hal bonds by independent methods (for example, the NQR method /54/, or from refraction and dielectric constant studies /55/). The result of the comparison, shown in Fig. 10, enables us to find the value of δ for an extrapolated 100 % ionic bond in SnX_4, giving $\delta = -(5.6 \pm 0.5)$ mm/sec $= -(4.4 \pm 0.4) \cdot 10^{-7}$ eV (which, by the way, is 2.5 mm/sec greater than the value of δ for SnF_4). Note also that even the point $\delta = 0$ for β-Sn itself fits beautifully on the general linear curve given in Fig. 10 for the chemical shift as a function of the electronegativity of X (the Sn--Sn bond in this case). Making use of the fact that for completely covalent and completely ionic quadrivalent tin bonds, the difference in $|\psi(0)|^2$ is determined by one 5s-electron (5 sp^3 hybridization, as we have seen, does not change matters), we obtain from (4):

$$\frac{\Delta R}{R} = \frac{(4.4 \pm 0.4) \cdot 10^{-7}}{1.55 \cdot 10^{-29} \cdot 1.5 \cdot 10^{26}} = (1.9 \pm 0.17) \cdot 10^{-4},$$

after which we can make a direct determination of the contribution to $|\psi_{5s}(0)|^2$ in the two tin compounds, from the value of the chemical shift. There is particular interest in all possible organic tin compounds, having strong quadrupole splitting, up to $\Delta = 3.6$

mm/sec (in general absent in all quadrivalent tin compounds except SnF_4), and having chemical shifts in the range (−1)--(−1.8) mm/sec, i.e., in the vicinity of the values of δ for SnR_4, where R is an organic radical.

A comparison between the values of quadrupole splitting for tin compounds and the Townes--Daley calculations /16/ of the field gradients q can so far only be made on the basis of a very rough approximation for the quadrupole moment of Sn^{119*}, $Q = 8 \cdot 10^{-26}$ cm^2 /28/. These calculations give a distance between peaks of $\Delta = 4.6$ mm/sec for a pure p_z electron, and $\Delta = 3.5$ X mm/sec, if, with complete sp^3-hybridization, one of the quadrivalent tin bonds is of a partly ionic character (the fraction x). With complete sp^3d^2-hybridization, appreciable quadrupole splitting is obtained for the configurations 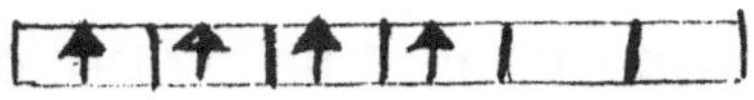 ($\Delta \gtrsim 4.6$ mm/sec) or  where one of the six bonds is completely ionic ($\Delta = 2.4$ mm/sec). In compounds of the type Ph_3SnHal, complete sp^3-hybridization may be assumed, with a partially ionic character of one of the bonds. With the value given above for $Q(Sn^{119*})$ we then get $X = 0.55$ (I), 0.7 (Br, Cl), and 1(F). It would be very interesting to make a direct determination of the effective charges of the tin and halogen atoms in these compounds, and thus find the quadrupole moment of Sn^{119*} directly. In order to explain the quadrupole splitting of the Mössbauer spectra of quadrivalent tin, we can evoke the idea of sp^3d^2-hybridization, i.e., that structures exist in these cases where the tin has coordination number six. Having quadrupole splitting of the Mössbauer spectra, for example, provides proof in favor of the idea that SnF_4 is an inorganic coordination polymer /56/, in which, for example, the hybridized

sp^3-electrons of the tin can give four, while the uninvolved electron pairs of the fluorine can give two more coordination bonds between the tin and the fluorine. Thus, the fluorine here forms bridging bonds between the tin atoms. The bridging structures (in particular, those with polycentric orbits), which, as has been pointed out by Ya. K. Syrkin /57/ are in general exceedingly wide spread, may in principle occur in a number of organic tin compounds. Since, however, it has been shown by structural analysis that a number of compounds of the type R_iSnHal_{4-i} (for example $(CH_3)_3SnI$, $(CH_3)_3SnCl$, and $(CH_3)_2SnCl_2$) have a tetrahedral configuration (see for example /58/), one must obviously in these cases look for an explanation of the quadrupole splitting within the framework of the four sp^3-bonds of the tin atoms.* In addition, a

---

*One more attempt at interpreting the form of the Mössbauer spectra for organic tin compounds (for example $C_4H_9)_2SnHal_2$) has been made in a new paper by A. Yu. Aleksandrov et.al. /ZhETF 43, 1242 (1962)/, experimental results of which are close to those in /29/. However, the assumption advanced in this new paper to the effect that "in $(C_4H_9)_2SnHal_2$ compounds, the bond between the tin atom and the halogen atom is formed mainly by p-electrons, while the s-electron part of the wave function is concentrated principally in the bond between the tin atoms and the C_4H_9 groups" is based on the purest lack of understanding. It is clear that no two s-electrons of any ns-shell can give two bonds without pairing off, and one of them passing to another state, otherwise, for example, the helium atom would be bivalent in the ground state.

molecular weight determination of Ph_3SnCl made in our experiments /34/ has shown that at least in this case we are dealing with a monomer.

Note in concluding this section, that it is precisely the organic tin compounds that have first made it possible to interpret the asymmetry and the doublet splitting of the Mössbauer spectra in isotropic polycrystalline samples /15,34,35/. We have observed strong asymmetry in the splitting for a number of derivatives of the type R_iSnHal_{4-i} (see for example Fig. 11), which persisted even when the test material was ground up or dissolved, hence could not be accounted for by the orientation of the samples. Magnetic suspension, and studies of the samples on an EPR outfit have shown that the asymmetry observed could not be due to ferromagnetic or paramagnetic impurities either. The asymmetry of the two quadrupole splitting peaks in the spectrum of a polycrystalline isotropic sample is produced by anisotropy of the Mössbauer effect in the single crystals themselves. As a matter of fact, as may be shown from /59/, if the axis of the axially symmetric electric field of a single crystal is oriented at an angle ϑ to the direction of the γ-quanta, the ratio of the π- to the σ-transition intensities is:

$$\frac{i_\pi(\vartheta)}{i_\sigma(\vartheta)} = \frac{2\sqrt{5}\,\overline{P}_0(\cos\vartheta) + \overline{P}_2(\cos\vartheta)}{2\sqrt{5}\,\overline{P}_0(\cos\vartheta) - \overline{P}_2(\cos\vartheta)} = \frac{1 + \cos^2\vartheta}{5/3 - \cos^2\vartheta}. \qquad (5)$$

If this ratio is averaged over the angles, which is equivalent to using an isotropic polycrystalline sample, we obtain

$$\frac{i_\pi\,(\text{total})}{i_\sigma\,(\text{total})} = \frac{\int i_\pi(\vartheta)\, d\cos\vartheta}{\int i_\sigma(\vartheta)\, d\cos\vartheta} = 1$$

However, if the Mössbauer effect is anisotropic, i.e., the probability of γ-transition without phonon emission itself depends on the angle ϑ, we have, in the form $f(\cos\vartheta)$

$$\frac{i_{\pi\,(\text{total})}}{i_{\sigma\,(\text{total})}} = \frac{\int_{-1}^{+1}(1+\cos^2\vartheta)\, f(\cos\vartheta)\, d\cos\vartheta}{\int_{-1}^{+1}\left(\frac{5}{3}-\cos^2\vartheta\right) f(\cos\vartheta)\, d\cos\vartheta} \neq 1, \qquad (6)$$

i.e., the asymmetry occurs in the doublet splitting. Here, neglecting fluctuations of the electric field gradients, the two peaks are of similar form (i.e., we have $y = f(x)$ for one peak, and $y = \text{const}\,[f(x)]$ for the other peak) both for uniaxial crystals, and for those cases where the crystalline field of the single crystal is not axially symmetric.

However, if we also take into consideration fluctuations in the gradients, the forms of the two peaks are different in addition to the percent absorption at the minimum. This question has been given a detailed theoretical treatment by S. V. Karyagin in /35/, while an experimental confirmation of the explanation given for the asymmetry has been provided in /34/ by comparing the spectra for Ph_3SnCl in the four different forms:

a) isotropic polycrystalline sample (at an angle of 90° to the γ- beam),

b) isotropic polycrystalline sample (at an angle of 45° to the γ-beam),

c) partly oriented sample (at an angle of 90° to the γ-beam),

d) partially oriented sample (at an angle of 45° to the γ-beam.

Changing the angle at which the sample is oriented to the γ-beam did not produce any change in the asymmetry of the peaks on going from a) to b), but did produce a change on going from c) to d). On the other hand, the partial orientation of the sample produced a change in the asymmetry of the peaks at both 90° and 45°. Grinding up the sample and destroying the partial orientation produced a change in the spectra observed in c) and d), and restored the picture found in a) ≡ b).

The fact that the results of a) and b) are identical eliminates any trivial explanation of the asymmetry of the two peaks involving quadrupole splitting due to the presence of anisotropy and hence of a definite orientation of the sample with respect to the γ-beam. The change in the spectra on going from b) to c) and then to d) removes any possibility of explaining the asymmetry as being the superposition of two singlet lines with different chemical shifts.

Thus, the results presented eliminate any need of interpreting the asymmetric doublet splitting as a superposition of two different chemical shifts, -- an interpretation, which, as we have seen at least in the case of $FeCl_3 \cdot 6H_2O$, leads to obvious misunderstandings. In addition, these results open up new possibilities of investigating the characteristics of single crystals by observing the Mössbauer spectra of polycrystalline samples. The local magnetic fields in the region where iron nuclei are located, as measured in a number of papers by means of the Mössbauer effect, also depend on the chemical state of the iron. In metallic iron /60/, cobalt, nickel /23/, and copper-nickel alloys /61/, the local magnetic fields at the iron nuclei are approximately the same, and close to 300 kOe. Accordingly, these fields are determined by the electron shell structure of the iron itself, they arise, for example, as a result of contact Fermi interaction produced by exchange polarization of the internal s-electrons of the unfilled 3d-shell of the iron atoms. In oxides containing trivalent iron, as was first shown in /10/, even stronger fields are observed, close to 500--550 kOe, and almost the same in all the systems of this sort that have been studied (see for example /41/), with the exception of the tetrahedral lattices of yttrium--iron garnets, where H = 390 kOe /40/. Thus, it is possible in these cases to speak of definite local magnetic fields, characteristic of the structure of the Fe″ electron shells themselves. Finally, in the ferrides of a number of rare earth elements (Sm, Gd, Dy, Ho, Er, Tm) of the type $SmFe_2$, the magnetic field at the iron nuclei is also almost the same and equal to ~ 230 kOe, except that in $CeFe_2$ a second value of H = 310 kOe /31/ is observed in addition. Thus, the electron configuration of the iron atoms is approximately the same even in all these compounds, and, in addition, judging from the chemical shift -- see Fig. 5, is nearly the same as that found in metallic iron, while the contribution from polarization of conductance electrons is very small, or is determined by interaction with the d-electrons of the iron. For CeF_2 it is assumed in /31/ that the 4f-electrons of the cerium are transferred to the d-band of the iron, without being localized at any definite iron atoms (judging from the chemical shifts). As a maverick among other ferrides, we find the uranium ferride, UFe_2 investigated in /32/, for which H = 20 kOe at room temperature and 65 kOe at -195° C. The least definite character so far is that shown by the data on the local magnetic fields at nuclei of bivalent iron, which apparently depend on the nature of the environment around the iron atoms, and are recounted in the paper by Wertheim /30/ already mentioned, from which Fig. 4 above was borrowed. For FeF_2, H = 340 kOe and for $Fe^{\cdot\cdot}$ in Fe_3O_4, H = 450 kOe. In the same paper /30/, where two forms of iron -- Fe^{2+} and Fe^{3+} -- were formed in beta decay of Co^{57} in cobalt oxide CoO, it was found that the field at the bivalent iron nuclei at very low temperatures (CoO is an antiferromagnetic cubic structure starting at the Neél temperature of + 18°C) is 200 kOe, while the characteristic value of H = 560 kOe is observed for $Fe^{\cdot\cdot\cdot}$ even here. Figure 4 above was obtained in /30/ at t° = 25° C, and thus corresponds with no Zeeman splitting.

In summing up this short review of the data on magnetic splitting of the Mössbauer spectra of iron, it may be said that in contrast with chemical shifts and quadrupole splitting, this data has not yet received any quantitative interpretation based on the electron shell structure of iron, and for this reason we have given them only a short treatment here. Many other elements have shown themselves to be good objects for Mössbauer effect studies, in addition to iron and tin. However, the basic problem of this work has so far been simply to show that the effect exists, and to investigate the principal nuclear characteristics of the γ-transitions (decay scheme, level width, etc.). Only in the case of gold (Au^{197*}) have studies already been made of the chemical shift changes as a function of the composition of the lattice (Au, Pt, steel, Fe, Co, and Ni), in which the gold was introduced in insignificant impurity quantities /62,63/. Accordingly, in this review, we are limiting ourselves to a detailed discussion of the results for iron and tin, mentioning the other elements -- in individual cases -- only in the following concluding section, which is devoted to some of the prospects of using the Mössbauer effect in chemistry.

V. SOME OF THE PROSPECTS FOR USING THE MÖSSBAUER EFFECT IN CHEMISTRY

Since the basic parameters of the Mössbauer spectra, such as the chemical shifts and the quadrupole splitting, are to a considerable degree determined by the valence electron shells of the Mössbauer atoms, a first and natural possibility of applying the Mössbauer effect to chemistry is to investigate the nature of the bonds formed by the atoms. Here, the simplest approach is to make a distinction between two types of bonds, ionic and covalent, and find the contribution made by each type. Without special calculations however, even this simplified approach is only suitable for explaining why different compounds have a certain relative position in a series of values of chemical shifts or quadrupole splitting. Any attempts at quantitative interpretation require making rather laborious calculations of the electron density $|\Psi(0)|^2$ and the electric field gradient $q = \partial^2 V/\partial Z^2$ in the region where the nucleus is located. With the appearance of rapid electronic computing machines, making such calculations has become a perfectly real thing, and it is necessary to make them.

In addition, it is very desirable to make any sorts of experiments that will permit an independent determination of such nuclear characteristics of Mössbauer emitters as the relative dimensional change $\Delta R/R$, and the quadrupole moment Q. Thus, for example, it would be a very necessary thing to measure the isotope shifts in the optical spectra of different compounds of Mössbauer atoms. The value of $\Delta R/R$ with change in mass number may be found with good accuracy, and thus from the values of the isotope shifts it would be possible to find the values of $|\psi(0)|^2$, and then use them to interpret the Mössbauer spectra. In finding the values of the quadrupole moments and the subsequent determination of the electric field gradients, it would be important to investigate Coulomb excitation of Mössbauer levels. As we know, the data on quadrupole splitting in conjunction with the simplified methods of calculating the gradients /16/ widely used at the present time still do not permit us to arrive at a unique conclusion as to the nature of the chemical bonds, since the reduction in the main contribution to the value of q, usually due to p_z-electrons, may be caused either by strengthening the ionic nature of the bonds, or by sp-hybridization of the valence bonds. However, the Mössbauer spectra have an additional parameter, the chemical shift, which sometimes makes it possible to decide between the above two possibilities. Thus, for example, in the quadrivalent tin already discussed, taking off two 5p-electrons ($5s^2p^3 \longrightarrow 5s^2$) increases the value of $|\psi(0)|^2$, while hybridization ($5s^2p^2 \longrightarrow 5sp^3$) acts in the opposite direction. The ability to distinguish between the effects of s- (chemical shifts), p- and d- (quadrupole splitting) electrons in the valence shells is an important advantage of the Mössbauer spectra, which, however, if it is to be taken advantage of completely, requires the development of a generalized $\delta\Delta$-method of calculation if only in the Townes--Dailey sense /16/. In addition, it must not be forgotten that dividing the chemical bonds into ionic and covalent is in itself a rather rough simplification, since no account is taken of the possibility of forming donor-acceptor and dative bonds, nor of bonds formed by polycentric orbits, nor of other forms of the chemical bond observed in recent years /50/. Not one of the methods now in existence for investigating electron structures, including the Mössbauer effect, can give an exhaustive answer as to the nature of the chemical bond, and there must be comparison and mutual complimentation of the information that they give, like the way this was demonstrated above in the case of the Mössbauer spectra and the effective charges of iron. The Mössbauer effect provides the simplest experimental method of observing even very small changes in the $|\psi(0)|^2$-value, which is relatively weakly dependent on the macroscopic structure of the sample, but very sensitive to screening of the Mössbauer nuclei by the electrons in the various molecular orbits. Accordingly, in addition to the studies already made on the way in which $|\psi(0)|^2$ depends on the nature of the neighbors of the Mössbauer atoms, a very promising problem is that of observing the changes in the chemical shifts as a function of the nature of the more remote chemical bonds. This outlook is particularly attractive for various organic tin compounds, in view of the fact that tin is an analog of carbon, and much of the information thus obtained can be made direct use of in organic chemistry. It would be interesting to introduce into organic tin molecules various groups showing a strong inductive or isomeric effect or large inductomeric polarizability, and then try to detect what effect they have on the electron density at the tin nuclei. Here we can compare the effect of the groups not merely as a function of how far they are from the tin atoms, but at the same time as a function of the nature of the intermediate bonds -- how, for example, the electron density is transmitted along a fatty chain, without interrupting a conjugated chain, and through an aromatic ring, in which the substituents are in the O--P or M position, etc. It is not impossible that the Mössbauer effect in tin-119 will be too "rough" an instrument ($\Gamma/E = 10^{-12}$) to observe the effect of remote chemical bonds in any given molecule, but there is no doubt that there is a general possibility in principle of recording the changes which they produce in the Mössbauer spectra.

As we have already mentioned above, the principal things that the Mössbauer effect can be applied to in chemistry are apparently metalloorganic and complex compounds. In the field of metalloorganic compounds there is substantial interest in making a comparison between the general nature of metallocarbon bonds, since there is a great difference between them for transition metals and metals belonging to the basic groups. Recently, A. N. Nesmeyanov gave a very complete review of the state of this problem /64/, on the basis of which we can easily lay out the main lines of application of the Mössbauer effect. Thus, the Mössbauer effect could be used to compare the acetylenidé complexes of transition metals, of the type $K_4Fe(C{\equiv}CH)_6$ (similar to the cyanide complexes, and formed apparently with the d-levels of Fe), with the alkenyl compounds of metals of the type of $Sn\text{-}(CH{=}CHCl)_4$, where there is an Me--C σ-bond (resulting from sp^3-hybridization). It would also be interesting to make a systematic comparison of the Mössbauer spectra of cyclopentadienylides of metals, $Me(C_5H_5)_2$, with various types of ferrocene-like "sandwich" structures, of the σ-derivatives of the type of dicyclopentadienyl tin, and of the linear as well as the "sandwich", but ionic structures of the type

$Me^{++}[(C_5H_5)^-]_2$. One of the interesting features, already observed by means of the Mössbauer effect, which is now undergoing careful study and explanation, is that there is so much similarity between the Mössbauer spectra for ferrocene and the phthalocyanine of bivalent iron. On the other hand, iron carbonyl, which has very nearly the same quadrupole splitting as ferrocene, shows a quite different chemical shift (the value of $|\psi(0)|^2$ for carbonyl is much higher--see Fig. 8). In order to make a comparison between the iron bonds in π-cyclopentadienyl compounds and those in carbonyls, it is desirable to investigate the Mössbauer spectra of mixed cyclopentadienylide carbonyls, polycarbonyls, and polyferrocenes. Related to carbonyls (and complex acetylenides) are also the complex compounds such as cyanide salts. There are also mixed carbonyl-cyanide derivatives with anions of the type $[Fe(CO)(CN)_5]'''$. Iron carbonyl, which has nearly the same chemical shift as the cyanide complexes, is nevertheless different in having many times larger quadrupole splitting. As we have seen above in the case of ferrocene, the quadrupole splitting of π-cyclopentadienylides yields easily to a semiquantitative treatment based on the calculations of /51,52/. It is accordingly a very good idea to investigate the changes in quadrupole splitting during a gradual transition from the carbonyl to the cyanide complexes--passing through the mixed forms, finally getting to heterometallic carbonyls, and carbonyls that are partially substituted by nitrogenous bases, etc. Among the metalloorganic compounds of transition elements, it would be interesting to make a comparison between the chemical shifts in the series SnR_4 (or SnR_iX_{4-i}), where R = aryl, alkyl, or vinyl. The thing is that the Me--C bond strengths are considerably less in alkyl derivatives, i.e., the nature of the bond is not completely determined by the difference in electronegativity between the metal and carbon, but depends to a large extent on the nature of the organic radical. Aryl and vinyl radicals are more easily removed in the form of anions than alkyl radicals, but it is more difficult to remove them in the form of radicals or cations. As a comparison standard, Mössbauer spectra should be taken of a hybrid of tin and mixed H_iSnR_{4-i} derivatives.

The particular interest in investigating the Mössbauer effect in iron is that iron, in one form or another, forms part of very many biologically important structures. It is true that the amount of iron in these structures is very small, so that for the investigations to succeed it would be necessary to synthesize compounds enriched in Fe^{57}, but even now the first results have been obtained in this field by using hemin. One of the problems which using the Mössbauer effect can make an important contribution to solving, is finding out whether or not, as shown by L. A. Blyumenfel'd's data /65/, there is anything specific about the structure of the electron shells and local magnetic fields at the iron nuclei occurring as small impurities in DNA and the DNA complexes formed with proteins.

In speaking of the problems of chemical structure, we must again mention the polymeric coordination compounds, based not on ordinary covalent bonds but on donor-acceptor, two-center, two electron bonds, as well as the various polycentric bonds (particularly electron deficit polymers). The bridge structures produced by these bonds are of such wide occurence that if the gross formulas of organic compounds represent their true structure, it must not be regarded as the rule but rather as a rare exception. For this reason, for example, Ya. K. Syrkin says /57/ that "the concept of trivalent iron is evidently out of date, since iron does not form three bonds". Thus, according to /57/, in $FeCl_3$, as a result of covalent and donor-acceptor bonds, each iron atom is bound with six chlorine atoms, while each chlorine atom forms bridging bonds between two or three iron atoms. As was shown above for $SnHal_4$, the Mössbauer effect enables us to draw a fairly sharp dividing line between bonds of the usual, two electron type (Hal = Cl, Br, I) and the formation of coordination structures (SnF_4) It is not impossible that a similar difference exists between $FeCl_3$, with $\Delta = 0$, and $FeCl_3 \cdot 6H_2O$, for which, apparently, $\Delta = 0.95$ mm/sec. Accordingly a systematic comparison must be made between the Mössbauer spectra and the chemical shifts and quadrupole splitting calculated under various assumptions as to the nature of the bonds. Compounds should also be studied, for which it is possible to have coordination polymer structures, for example, the tin alkoxydes, in which it is to be expected that the Mössbauer spectrum will be similar to SnF_4. A number of coordination polymers may be formed in solutions, such, for example, as mono-, di-, and trimeric complexes, $SnOH^+$, $Sn_2OH_4^{2+}$, and $Sn_3(OH)_4^{2+}$, which occur in the hydrolysis of Sn_2^+, and the complexes $FeOH^{++}$, $Fe(OH)_2^+$, and $Fe_2(OH)_2^{4+}$ occurring in the hydrolysis of Fe_3^+ /56/. The formation of these complexes will also affect the Mössbauer spectra, and in this connection we immediately encounter possibilities of investigating both structural and kinetic problems (the properties of the intermediate products), to which we shall soon turn. For the present however we shall point out the particular interest that there would be for the chemistry of complex compounds in observing and making use of the Mössbauer effect in platinum. We shall give one example here of a quite well-known coordination polymer, trimethylchlorplatinum, $Pt(CH_3)_3Cl$. It is analogous in structure to tetramethylplatinum, where polycentric orbits are used to form the tetrameric molecules shown in Fig. 12a. A similar tetramer may be prepared from trimethylchlorplatinum. However, as may be seen from Fig. 12, this tetramer can in principle exist in two forms $[Pt(CH_3)_3Cl]_4$ (Fig. 12b), in which the bridging bonds are formed by chlorine atoms, and each platinum atom is bound directly to three chlorine atoms and three methyl radicals, and $\{Pt[(CH_3)Cl]CH_3\}_4$ (Fig. 12c), where the bridging bonds are formed by methyl radicals, and each platinum atom is bound to only one chlorine atom and five methyl radicals. This peculiar bridging isomerism would correspond to having two different chemical shifts in the Mössbauer spectra of platinum. Thus, the Mössbauer effect could in principle even help to observe the existence of new chemical phenomena.

In addition to problems in the structure of chemical compounds, the Mössbauer effect can, beyond doubt, also be used in chemical kinetics and radiation chemistry. In addition to the possibility of taking kinetic curves directly, all in one experiment (from the frequency of the readings at some fixed characteristic rate of motion), there are particularly interesting prospects here of observing unstable intermediate products. When reactions are being carried out in the liquid phase, we have to stop the process and freeze the mixture for each observation of the Mössbauer spectrum. However, with topochemical processes (in particular, and what is particularly interesting, -- radiation topochemical processes), the changes in the Mössbauer spectrum may be observed continuously during the reaction. An example of the possibility is provided by our observation of a sudden change in the form of the Mössbauer spectra while a solid equimolecular mixture of $SnPh_4 + SnI_4$ was being irradiated with 1.6 MeV electrons /15/. The change came from the formation of various disproportionation reaction products of the type Ph_iSnI_{4-i}. Among other radiation chemical reactions which may be investigated with the aid of the Mössbauer effect, we can mention the very peculiar radiation synthesis of organic tin compounds performed by K. A. Kocheshkov et.al. /66/ by irradiating tin in halogen substituted hydrocarbon derivatives:

$$Sn + 2R\text{Гал} \longrightarrow R_2 Sn\,\text{Гал}_2$$

as well as the radiation oxidation of bivalent tin by trivalent iron salts in aqueous sulfuric acid solutions, during which unstable intermediate products are assumed to be formed, which are derivatives of mono- and trivalent tin /67/. It is also possible in various solvents to observe heterolytic exchange of the type $SnR_4 + KCl \longrightarrow SnR_3Cl + KR$, in which the radical plays the role of an anion, and homolytic exchange of the type $Me + Me'R$, where the radical goes from one metal to another. It would be extremely interesting to use the Mössbauer effect to observe polymerization (for example, of monomers containing tin) in the solid phase, since in experiments on radiation solid phase polymerization we usually do not know whether the polymer is formed during the actual irradiation or during the subsequent thawing, for example, at the phase transition or melting points. Using the Mössbauer effect in polymerization studies is also promising for the reason that at the present time wider and wider use is being made as efficient polymerization catalysts of bimetalloorganic complexes containing tin, of the type of the heterogeneous catalyst $Sn(C_4H_9)_4 + TiCl_4 + AlCl_3$ /68/ (or of the homogeneous catalyst $Sn(C_6H_5)_4 + (C_5H_5)_2VCl_2 + AlBr_3$ /69/. Tetrabutyl tin has no appreciable ability to reduce $TiCl_4$, but it it is a good alkyl donor, and, when it reacts with the electrophilic $AlCl_3$, it forms a high activity catalyst. The problems of the state of the catalytic complexes, and of the change in electrical structure of the tin atoms during the actual polymerization will undoubtedly receive some additional light when Mössbauer spectra are taken. By putting Mössbauer atoms (principally tin) into polymer molecules, we can count on getting extremely interesting information on how the recoil energy is transferred to collective forms of excitation in these molecules. In particular, in working with molecules of very large molecular weight (more than 10^8 in the case of tin), such, for example, as cross link polymers, we can find out whether the recoil is taken up by the molecule as a whole, or whether it excites deformational vibrations. We shall also mention here the possibility already touched on above in connection with /30/, of using the Mössbauer effect to investigate various chemical consequences of nuclear transformations. It will be interesting in particular to see whether or not there are any sharp changes in the spectra when the Mössbauer atoms get into a zone close to the tracks of fission fragments or of the strongly ionizing particles produced when thermal neutrons are captured by

B^{10} and Li^6 nuclei, for the reason that in this zone (100 A in diameter) very strong short time (10^{-10} sec) heating occurs, followed by rapid cooling. Finally, attention should be given to the highly probable analytical applications of the Mössbauer effect, for example, in cases where all kinds of disproportionation reactions (of the type $SnR_4 + SnX_4 \longrightarrow RiSnX_{4-i}$) can occur to form products, which have substantially different Mössbauer spectra from the starting materials. The examples given could easily be multiplied, but there are probably enough already to give some idea of the abundant possibilities that exist in chemistry of making use of the Mössbauer effect, that discovery of nuclear physics, so remarkable for its simplicity and ingenuity.

Institute of Chemical Physics, Academy of Sciences, USSR

Translated by: Charles V. Larrick

REFERENCES

/1/ G. Breit, Rev. Mod. Phys 30, 507 (1958) and A. R. Striganov and Yu. P. Dontsov, UFN, 55, 315 (1955

/2/ R. Weiner, Phys. Rev. 114, 256 (1959), A. C. Melissinos and S. P. Davis, Phys. Rev. 115, 130, (1959)

/3/ C. Townes and A. Shavlov "Radiospectroscopy" /Russian translation/ Inoizdat 1959; K. Khausser, Uspekhi khimii 27, 403, (1958); N. M. Aleksandrov, F. I. Skrypov, UFN, 75, 585 (1961)

/4/ C. Towns and A. Shavlov "Radiospectroscopy" /Russian translation/ Inoizdat 1959; U. Orvill-Tomas, Uspekhi khimii 27, 731 (1958); V. Grechyshkin, UFN 69, 189 (1959); G. K. Semin and É. I. Fedin, Zhurnal strukt. khimii 1, 252 and 464 (1960)

/5/ E. Segre, Phys. Rev. 71, 274 (1947; E. Segre, C. Wiegand, Phys. Rev. 75, 39 (1949)

/6/ C. A. Accardo, Phys. Rev. Lett. 1, 180 (1958)

/7/ K. Bainbridge, M. Goldhaber and E. Wilson, Phys. Rev. 84, 1260 (1951); J. C. Slater, Phys. Rev. 84, 1261 (1951)

/8/ R. L. Mossbauer, Z. Physik, 151, 124 (1958) and Naturwiss. 45, 538 (1958)

/9/ A. R. Bodmer, Nucl. Phys. 21, 347 (1961)

/10/ R. V. Pound and G. A. Rebka, Phys. Rev. Lett. 4, 474 (1960); B. D. Josephson, Phys. Rev. Lett. 4, 341 (1960)

/11/ O. C. Kistner and A. W. Sunyar, Phys. Rev. Lett. 4, 412 (1960)

/12/ V. S. Shpinel', V. A. Bryukhanov, N. N. Delyagin, ZhÉTF, 41, 1767 (1961)

/13/ R. E. Watson, Phys. Rev. 119, 1934 (1960); R. E. Watson and R. I. Freeman, Phys. Rev. 120, 1152 (1960)

/14/ L. R. Walker, G. K. Wertheim and V. Jaccarino, Phys. Rev. Lett. 6, 98 (1961)

/15/ V. I. Gol'danskii, G. M. Gorodinskii, S. V. Karyagin, L. A. Korytko, L. M. Krizhanskii, E. F. Makarov, I. P. Suzdalev, V. V. Khrapov, DAN 147 (1962) No. 1

/16/ C. H. Townes and B. P. Dailey, J. Chem. Phys. 17, 782 (1949)

/17/ P. P. Craig, D. E. Nagle, D. R. F. Cochran, Phys. Rev. Lett. 4, 561 (1960); S. I. Aksenov, V. P. Alfimenkov, V. I. Lushikov, Yu. M. Ostanevich, F. L. Shapiro, Yang-Wu-Kiang, ZhÉTF 40, 88 (1961)

/18/ G. N. Belozerskii and Yu. A. Nemilov, UFN 72, 433 (1960)

/19/ H. Frauenfelder, Introductory paper in the collection "The Mossbauer Effect". W. A. Benjamin, Inc., New York, 1962

/20/ F. L. Shapiro, UFN 72, 685 (1961)

/21/ Yu. M. Kagan ZhÉTF 41, 659 (1961)

/22/ Yu. M. Kagan and V. A. Maslov ZhÉTF 41, 1296 (1961)

/23/ G. K. Wertheim, Phys. Rev. Lett. 4, 403 (1960)

/24/ S. L. Ruby, L. M. Epstein and K. H. Sun, Rev. Sci. Instr. 31 580 (1960)

/25/ V. V. Sklyarevskii, B. N. Samoilov, E. P. Stepanov, ZhÉTF 40, 1874 (1961)

/26/ N. N. Delyagin, V. S. Shpinel'. V. A. Bryukhanov, ZhÉTF 41, 1374 (1961)

/27/ V. A. Bryukhanov, V. I. Gol'danskii, N. N. Delyagin, E. F. Makarov, V. S. Shpinel' ZhÉTF 42, 637 (1962)

/28/ A. J. F. Boyle, D. S. P. Bunbury, C. Edwards, Proc. Phys. Soc. 79, 416 (1962)

/29/ V. A. Bryukhanov, V. I. Gol'danskii, N. N. Delyagin, L. A. Korytko, E. F. Makarov, I. P. Suzdalev, V. S. Shpinel' ZhÉTF 43, 448 (1962)

/30/ G. K. Wertheim, Phys. Rev. 124, 764 (1961)

/31/ G. K. Wertheim and J. H. Wernick, Phys, Rev. 125, 1937 (1962)

/32/ S. Komura, N. Kunitomi, P. Tseng, N. Shikazono and H. Takekoshi J. Phys. Soc. Japan 16, 1479 (1961)

/33/ E. A. Friedman and W. J. Nicholson, Bull. Am. Phys. Soc. 7 No. 6, 402 (1962)

/34/ V. I. Gol'danskii, S. V. Karyagin, E. F. Makarov, V. V. Khrapov, Transactions of the Conference on the Mossbauer Effect

/in Russian/ (Dubna, July 1962)
/35/ S. V. Karyagin, DAN (1962) (in press)
/36/ Yu. M. Kagan DAN 140, 794 (1961)
/37/ N. E. Alekseevskii, Pham Zuy Xien, V. G. Shapiro, V. S. Shpinel' ZhÉTF 43, 790 (1962)
/38/ S. de Benedetti, G. Lang and R. Ingalls, Phys. Rev. Lett. 6 60 (1961)
/39/ W. Kerler and W. Neuwirth, Z. Phys 167, 176 (1962); W. Kerler, Z. Phys. 167, 194 (1962)
/40/ C. Alff and G. K. Wertheim, Phys. Rev. 122, 1415 (1961)
/41/ G. Shirane, D. Cox, S. L. Ruby, Phys. Rev. 125, 1158 (1962)
/42/ G. Shirane, W. J. Takei, S. L. Ruby, Phys. Rev. 126, 49 (1962)
/43/ U. Zahn, P. Kienle and H. Eicher, Z. Phys, 166, 220 (1962)
/44/ L. M. Epstein, J. Chem. Phys, 36, 2731 (1962)
/45/ I. Solomon, Comp. Rend. 250, 3828 (1960); 251, 2675 (1960
/46/ E. Fermi and E. Segrè, Z. Phys. 82, 729 (1933)
/47/ S. A. Goudsmit, Phys. Rev. 43, 636 (1933)
/48/ É. E. Vainshtein, R. L. Barinskii, K. I. Narbutt, ZhÉTF 23 593 (1952)
/49/ R. L. Barinskii, Zhurnal struktur, khimii 1, 200 (1960)
/50/ Ya. K. Syrkin, Uspekhi khimii 31, 397 (1962)
/51/ E. M. Shustorovich and M. E. Dyatkina, DAN 128, 1234 (1959)
/52/ E. M. Shustorovich and M. E. Dyatkina, DAN, 133, 141 (1960)
/53/ V. A. Bryukhanov, N. N. Delyagin, Opalenko, V. S. Shpinel' ZhÉTF 43, 432 (1962)
/54/ A. L. Schawlow, J. Chem. Phys. 22, 1911 (1954)
/55/ M. M. Yakshin, V. M. Ezuchevskaya, V. A. Salmenkova, Zhurnal neorg. khimii 6, 2425 (1961)

/56/ I. Khaiduk, Uspekhi khimii 30, 1124 (1961)

/57/ Ya. K. Syrkin, Uspekhi khimii 28, 903 (1959)

/58/ I. P. Gol'dshtein, E. N. Gur'yanova, E. D. Delinskaya, K. A. Kocheshkov, DAN 136, 1079 (1961)

/59/ A. M. Baldin, V. I. Gol'danskii, I. L. Rozental', "Kinetics of Nuclear Reactions" /in Russian/ Fizmatizdat 1959

/60/ S. S. Hanna, I. Heberle, C. Littlejoh, G. J. Perlow, R. S. Preston, D. H. Vincent, Phys, Rev. Lett. 4, 177 (1960)

/61/ G. K. Wertheim and J. H. Wernick, Phys. Rev. 123, 755 (1961)

/62/ D. A. Shirley, M. Kaplan, P. Axel, Phys. Rev. 123, 816 (1961)

/63/ D. A. Shirley, Phys. Rev. 124, 354 (1961)

/64/ A. N. Nesmeyanov "D. I. Mendeleev's Periodic System of the Elements and Organic Chemistry" /in Russian/ Paper presented at the VIII Mendeleev Congress on General and Applied Chemistry, Moscow, (1959)

/65/ L. A. Blyumenfel'd, V. A. Benderskii, A. É. Kalmanson, Biofisika 6, 631 (1961); L. A. Blyumenfel'd, DAN 147 (in press) 1962

/66/ L. V. Abramova, N. I. Sheverdina and K. A. Kocheshkov, DAN 123, 681 (1958)

/67/ J. W. Boyle, S. Weiner, C. J. Hochanadel, J. Phys. Chem. 63 892 (1959)

/68/ A. V. Topchiev, B. A. Krentsel', L. L. Stotskaya, Uspekhi khimii 30, 462 (1961)

/69/ G. A. Abakumov, S. V. Shulyndin and A. E. Shilov, Kinetika i kataliz (in press) (1962)

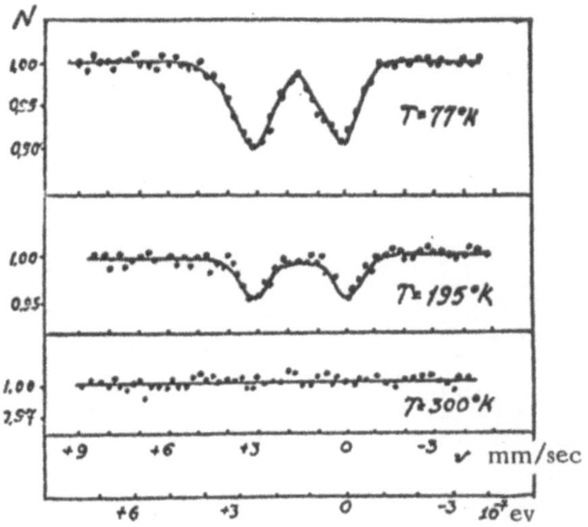

Fig. 1. Mössbauer spectra for polymer containing tin (on methylmethacrylate base) at different temperatures. Source--SnO_2.

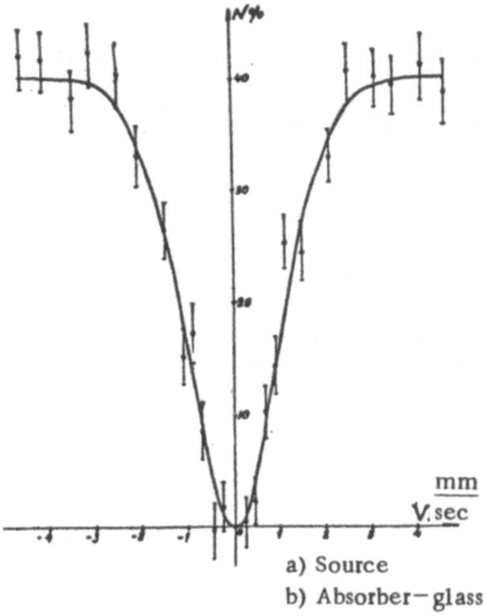

Fig. 2. Mössbauer spectrum for a glass containing tin (9.1% SnO_2) at 77° K. Source--SnO_2.

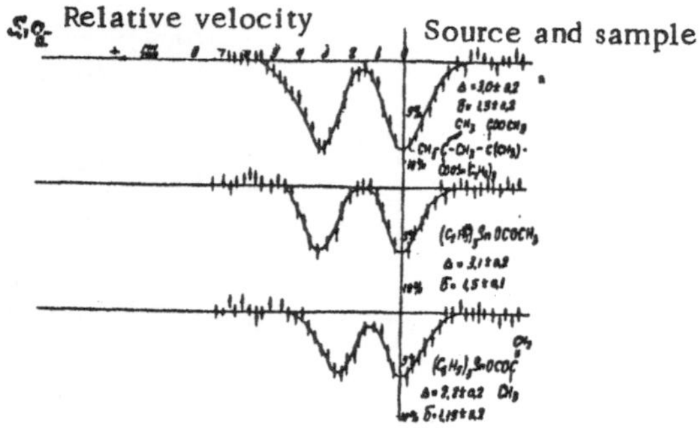

Fig. 3. Mössbauer spectra for two tin derivatives having the same chemical bonds nearest the tin: $(C_2H_5)_3Sn{-}{-}O{-}{-}C{\genfrac{}{}{0pt}{}{\not{}O}{\diagdown}}$

Lower spectrum--for an organic tin compound where the tin is bound to phenyl instead of ethyl radicals: $(C_6H_5)_3Sn{-}{-}O{-}{-}C{\genfrac{}{}{0pt}{}{\not{}O}{\diagdown}}$

Temperature 77° K. Source--SnO_2.

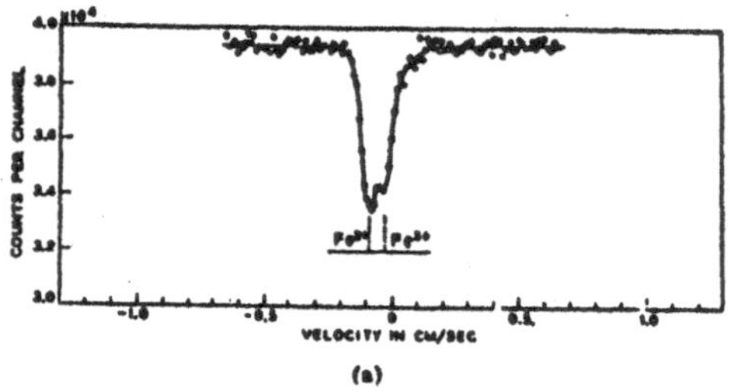

Fig. 4. Splitting of Mössbauer spectrum resulting from two chemical states of iron ($Fe^{''}$ and $Fe^{'''}$) being formed in beta-decay of cobalt 57 in cobalt oxide. Absorber--$K_3Fe(CN)_6$. Temperature 298° K.

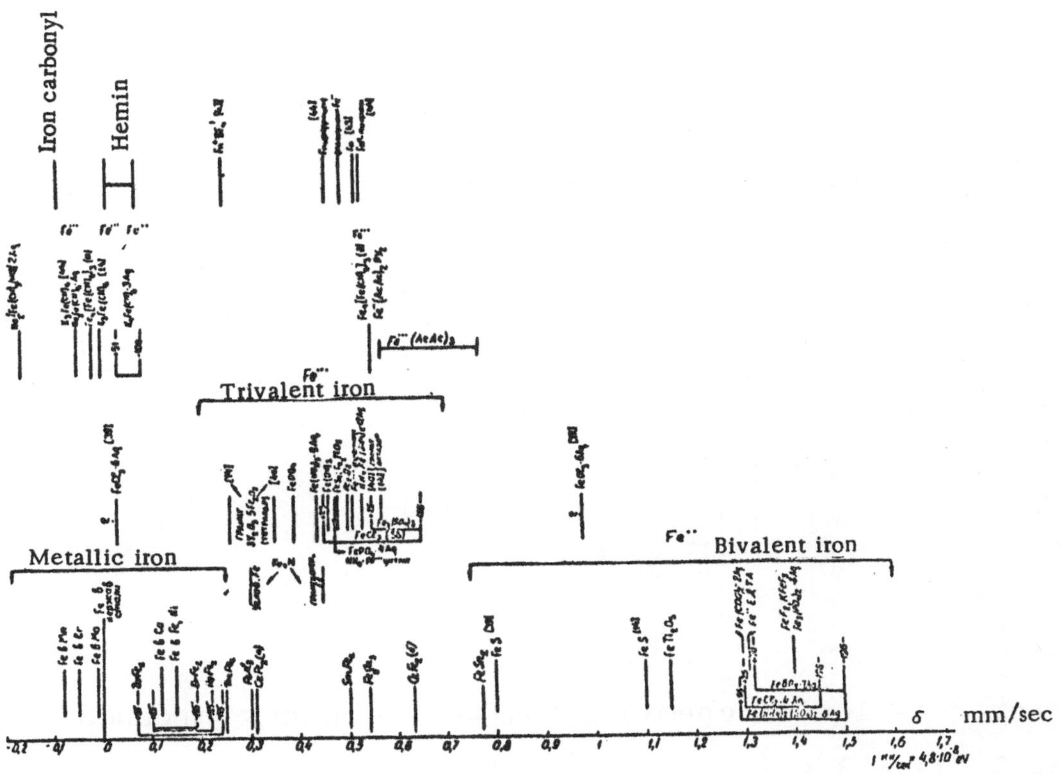

Fig. 5. Data on chemical shifts in the Mössbauer spectra of iron compounds, The zero position of the line is that for iron in stainless steel.

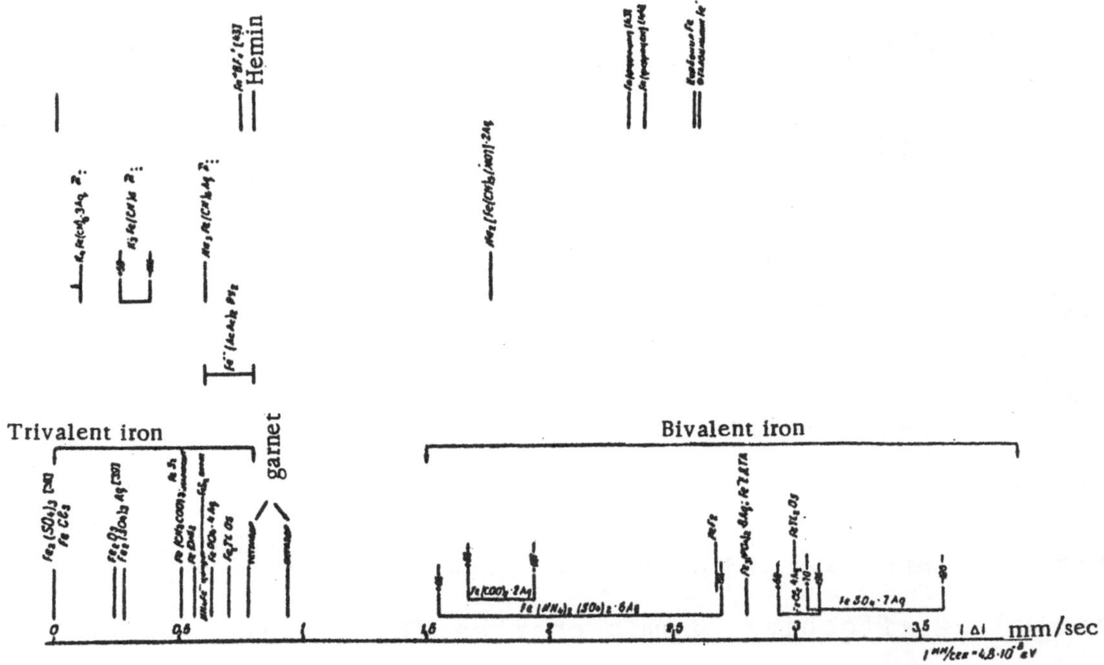

Fig. 6. Data on quadrupole splitting of the iron line in the Mössbauer spectra of iron compounds.

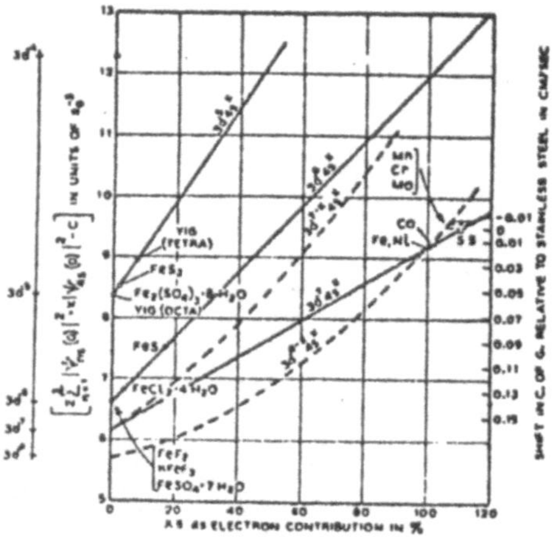

Fig. 7. Interpretation of chemical shifts for iron compounds /14/. The constant C =11.873 a_0^{-3}.

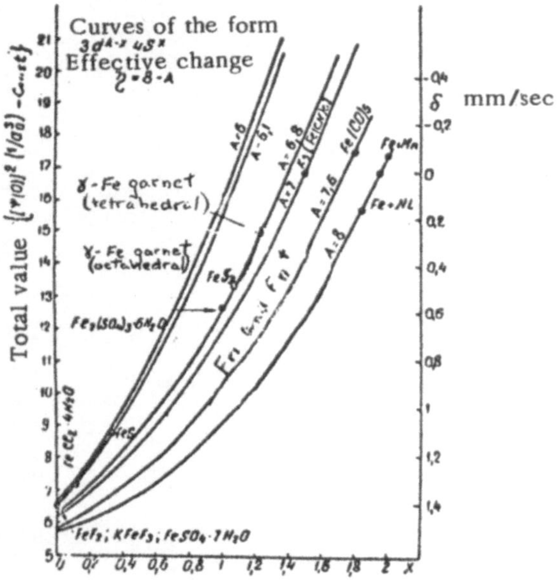

Fig. 8. Alternate interpretation of the chemical shifts for iron compounds, including effective atomic charges (but neglecting sp^3 hybridization for complex compounds). The constant (ordinate axis) is equal, as in Fig. 7, to 11.873 a_0^{-3}.

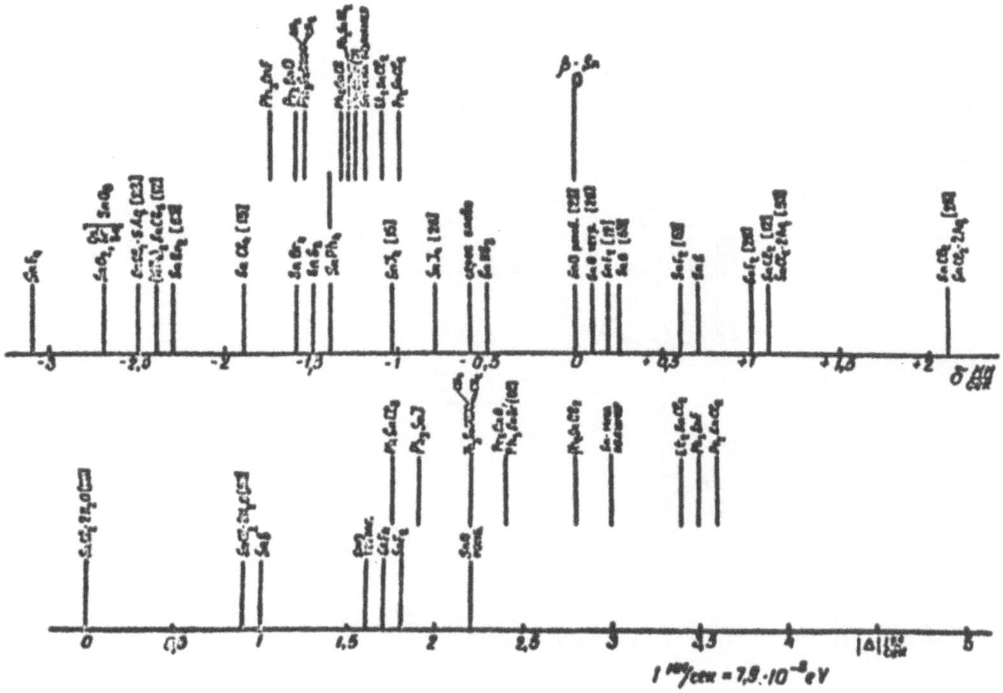

Fig. 9. Data on chemical shift (with respect to β-Sn) and quadrupole line splitting in the Mössbauer spectra of tin compounds.

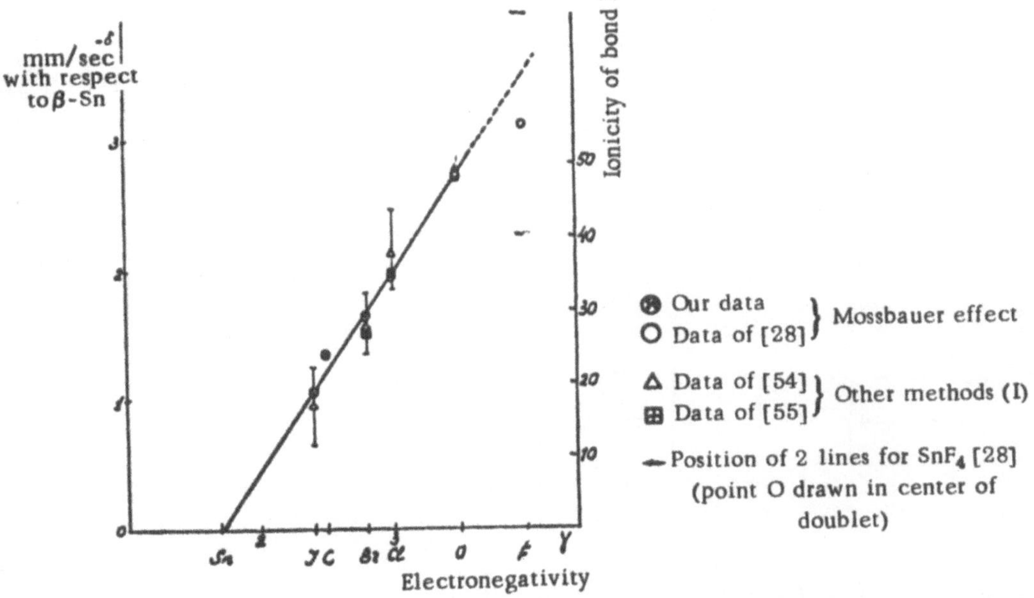

Fig. 10. Chemical shifts in SnX_4 compounds as a function of the electronegativity X and degree of ionicity of the Sn--X bonds.

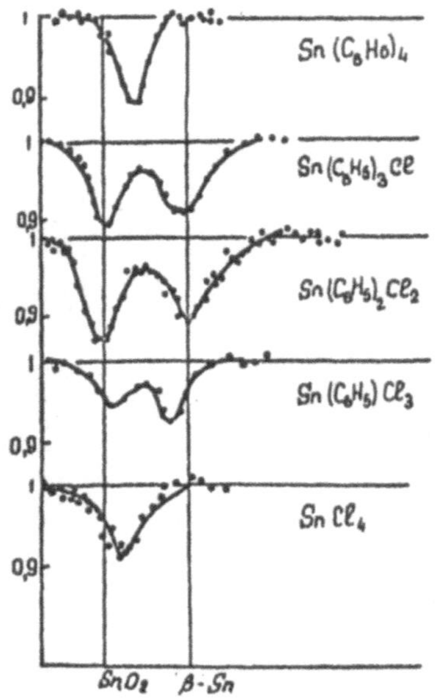

Fig. 11. Illustrating asymmetry of the doublet splitting of Mössbauer spectrum lines in organic tin compounds Ph_iSnCl_{4-i}.

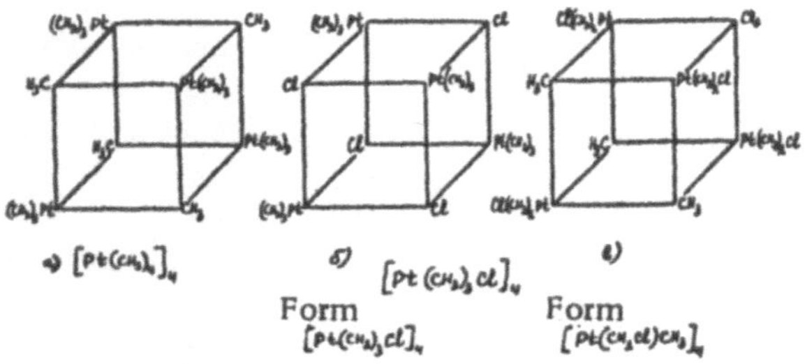

Fig. 12. Illustrating possible bridge stereoisomerism for trimethylchlorplatinum tetramer

THE FINE STRUCTURE OF MÖSSBAUER SPECTRA IN MINERAL COMPOUNDS OF IRON

A. Z. Grinkevich, D. S. Kul'gavchik

The Mössbauer effect was used to determine the resonance absorption of 14.4 keV γ-rays by the nucleus of Fe^{57} in mineral compounds of iron.

The results of the investigation of the fine structure of γ-transition are given in the table.

Compound	Crystal structure	Measured displacement	Quadrupole splitting		H_{eff}, Oe.
			mm/sec	MHz.	
$FeCO_3$ Siderite	Trigonal	−1,30±0,05	1,80±0,05	10,5±0,3	0
γ−FeOOH Lepidocrocite	Rhombohedral	−0,43±0,05	0,55±0,05	3,2±0,3	0
α−FeOOH Goethite	Rhombohedral	−0,67±0,05	0,53±0,05	3,1±0,3	$(305\pm35)\cdot10^3$
$NaFe_3(SO_4)_2(OH)_6$ Jarosite	Trigonal	−0,48±0,05	1,00±0,05	5,8±0,3	0
$KFe_3(SO_4)$ Jarosite	Rhombohedral	−0,45±0,05	1,10±0,05	6,4±0,3	0

Institute of Nuclear Physics, Crakow

Institute of Physics, Jagiellonian University, Crakow

ANISOTROPY OF THE MOSSBAUER EFFECT IN A SINGLE CRYSTAL OF β-Sn

N.E. Alekseevskii, Pham Zuy Xien, V.G. Shapiro, V.S. Shpinel'

ANNOTATION

The probability of resonance absorption f' of the 23.8 KeV γ-quanta was measured in monocrystalline platelets of white metallic tin, cut out in various crystallographic planes. A large anisotropy effect was found which did not change on going from a temperature of 293°K to a temperature of 77°K. The ratio of the f' values obtained in the different platelets is $f'_{100} : f'_{101} : f'_{001} : f'_{polycryst.} = 1 : 0.95 : 0.71 : 0.80$. The differences found in the positions of the absorption maxima and the asymmetry of the absorption lines is explained by the quadrupole splitting of the excited level of the Sn^{119} nucleus in the β-Sn crystal.

1. INTRODUCTION

Starting from qualitative considerations, it can be expected that the probability of emission (or absorption) of γ-quanta without recoil should depend on the orientation of the crystallographic axes with respect to the direction of the flux of γ-quanta.

The calculation carried out in /1/ shows that, in the tetragonal lattice of white tin, the probability of the Mossbauer effect along the c axis at T = 0°K is 4% greater than in a perpendicular direction.

The object of the present work was the investigation of the anticipated anisotropy in a single crystal of white tin.

2. DESCRIPTION OF THE EXPERIMENT

The measurements of resonance absorption were carried out at room temperature and at liquid nitrogen temperature with absorbers in the shape of platelets, cut from a single crystal of natural white tin (β-Sn) in the (001), (101) and (100) planes. Polycrystalline tin foil was used in the control measurements. The thicknesses of these absorbers, as determined from measurements of their non-resonance absorption of γ-rays at room temperature, were in the ratio 1: 1.01: 0.95: 0.94, respectively, where the thickness of the first, as determined by weighing, amounted to 57.5 ± 0.8 mg/cm^2. In measurements at liquid nitrogen temperature, the relative thicknesses of the platelets cut out in the (100) and (001) planes and of the polycrystalline foil were respectively 1: 1.15: 1.08, where the thickness of the first was 20.0 ± 0.2 mg/cm^2.

The Sn^{119m} source of 23.8 KeV γ-rays was prepared from tin dioxide, SnO_2, enriched in the Sn^{118} isotope to 88% and irradiated in a reactor. The thickness of the source amounted to approximately 5 mg/cm^2. The source was at room temperature for all the measurements.

The beam of γ-quanta was cut by a lead diaphragm with an aperture of 6 mm diameter. The absorbers studied were set up beyond the diaphragm in a strictly fixed position. The radiation which passed through the absorber was detected by a luminescence spectrometer with a plane NaI(Tl) crystal.

The measurements were carried out using apparatus described earlier /2/, permitting the absorber to be set in motion with a constant velocity.

3. RESULTS OF THE MEASUREMENTS AND DISCUSSION

Some of the absorption spectra, taken on different platelets under identical geometrical conditions at a temperature of 77°K, are shown in the figure. As is seen from the figure, the absorption lines are situated in the positive velocity range. The counting level in the negative velocity range depends on the non-resonance absorption. The disposition of the experimental points in this region characterizes the stability of the recording apparatus.

The quantities which characterize the experimental curves are presented in Table 1: V_{res} — the position of the absorption maximum and $\mathcal{E}_{max}$ — the ratio of the difference in counting rates in the absence of resonance absorption and at the maximum of resonance absorption to the counting rate without resonance absorption.

The observed halfwidths of the lines exceed the halfwidths expected in this case taking account of the effective thicknesses of the absorber and of the source, C_A $(C_A = n_A f' \mathcal{E}_0)$ and C_s, respectively. For example, at a temperature of 77°K the halfwidth of the line for the polycrystalline foil amounts to 1.5 mm/sec, while it should be only 0.9 mm/sec in all. Vibrations which arise during the relative motion of absorber and source, the level of which was specially measured, can cause the line to broaden by no more than 8%. The observed broadening, though, can be due to quadrupole interaction. This circumstance must be taken into account in the determination of the probability of absorption of γ-quanta without recoil for the differently oriented platelets.

In this case it is convenient to use the method proposed in /3/ in the processing of the results of the measurements. For the determination of the amount of quadrupole splitting Δ and f', we use the analytical formulas obtained there for S_{exp} (the area of the absorption curve) and $\mathcal{E}_{max}$ (the relative maximum absorption; $\mathcal{E}\left(\frac{\Delta}{2}\right)$ is denoted in /3/), which depend on Δ and the effective thickness of the absorber, C_A.

In the work referred to, it is shown that determination of f' through measurement of the area of the resonance absorption curve is convenient inasmuch as S_{exp} does not depend on peculiarities of the emission spectrum of the source which arise as a consequence of self absorption, splitting of the emission line and vibrations.

For the calculation, one has to know such characteristics of the source as αf (where f is the probability of emission of γ-quanta without recoil, α is a parameter which determines the relative contributions of the γ-quanta studied to the total counting rate) and the effective thickness of the source $C_s = n_s f \sigma_0$ (where n_s is the number of resonance absorbing nuclei per cm^2 of the source). By means of measurements of the resonance absorption of γ-quanta of SnO_2 absorbers of various thicknesses (from 4.5 mg/cm^2 to 60 mg/cm^2) it was found that $\alpha f = 0.28 \pm 0.03$, $C_s \approx 0.4$.

Using the family of curves S_{exp} and $\mathcal{E}_{max}$ (Figs. 2 and 3 of /3/), we can determine Δ and f' for polycrystalline white tin (Table 2). The values obtained for Δ are in agreement with the data of /4/ within the limits of experimental error, if one takes the width of the line emitted by the source to be $\Gamma = 0.31$ mm/sec.

Now let us go to the discussion of the results for the single crystal. In a first approximation, one can assume that the gradient of the electric field in the β-Sn crystal has axial symmetry. The β-Sn crystal belongs to the tetragonal system, therefore there is every reason to assume that the axis of symmetry of the field gradient coincides with the c-axis of the crystal, i.e. it is perpendicular to the (001) plane.

As a consequence of the quadrupole interaction, the 23.8 KeV excited level of the Sn^{119} nucleus is split into two sublevels with spin projections $\pm 3/2$ and $\pm 1/2$ on the axis of symmetry.

By means of a kinematic calculation, one can establish that the probabilities of resonance absorption of γ-quanta via these sublevels are in general not equal, and they depend on the angle θ between the axis of symmetry of the crystal and the direction of the incident γ-quanta:

$$W_{\pm 3/2} \sim (1 + \cos^2\theta)$$

$$W_{\pm 1/2} \sim (5/2 - \cos^2\theta), \qquad (1)$$

where $W_{\pm y/z}$ is the probability of resonance absorption of γ-quanta via the corresponding sublevel.

Since the direction of the incident γ-quanta is perpendicular to the plane of the absorber, using the platelets cut out of a single crystal in different planes, we obtain the following values of relative intensity of the components of the absorption spectrum $a = W_{\pm 3/2}/W_{\pm 1/2}$:

Orientation of the platelet	a
(001)	3
(101)	9/7
(100)	3/5

(2)

Thus, the shape of the absorption line should depend on the orientation of the platelet, namely: the position of the absorption maximum shifts in the direction of the more intense component and the line becomes asymmetrical. The features indicated should be more clearly pronounced the greater is Δ and the smaller the effective thickness of the absorber. At room temperature, the asymmetry of the lines and the shift of the absorption maxima were not observed within the limits of experimental error. At a temperature of 77°K, as is seen from the figure, the absorption lines for the (001) and (100) platelets actually are asymmetric, more noticeably in the first case. The observed asymmetry indicates (taking (2) into account) that the sublevel with spin projection $\pm 1/2$ has a higher energy than the sublevel with spin projection $\pm 3/2$. From this, it follows that the product Qq_{zz} (where Q is the observed quadrupole moment of the nucleus and q_{zz} is the component of the gradient of the field along the axis of symmetry) should carry a minus sign.

For the level we are considering, the quadrupole splitting is $\Delta = (1/2)eQq_{zz}$. It is a quantity characteristic of the elementary cell of a given crystal. Therefore we can use the value of Δ obtained on a polycrystal to calculate f' in the platelets of different orientations.

We determined f' taking account of (2) and using the general formula (30) of /3/, (see Table 3).

From formula (30) it follows that, for a given value of the effective thickness of the absorber, the area of the absorption curve depends weakly on the relative intensity of the components and on Δ. Therefore possible errors in the determination of Δ, and also small deviations of the field gradient from axial symmetry, have little effect on the obtained results.

We note that the ratio of the f' values for absorbers of different orientations is determined more accurately than these values themselves, since it is almost independent of the results of measurements of other quantities $(\Delta, \alpha f)$.

For a temperature of 77°K, $f'_{100} : f'_{001} : f'_{polycryst.} = 1 : 0.67 : 0.89$;
For a temperature of 293°K, $f'_{100} : f'_{101} : f'_{001} : f'_{polycr.} = 1 : 0.95 : 0.71 : 0.80$.

Within the limits of experimental error, the ratios given do not change on going from liquid nitrogen temperature to room temperature.

Thus, from the experiment there follows a relatively smaller probability of absorption of γ-quanta without recoil when they propagate along the c axis. The relatively large probability theoretically predicted for the effect in this case (for T = 0°K) /1/ is difficult to reconcile with our results.

Institute of Nuclear Physics, Moscow State University

Table 1

Orientation of the platelet	77°K		293°K	
	V_{res} mm/sec 1)	$\mathcal{E}_{max}$ %	V_{res} mm/sec	$\mathcal{E}_{max}$ %
Polycrystal	2.53 ± 0.07	14.0 ±0.5	2.50 ±0.10 2)	8.4±0.5
001	2.33 ± 0.07	12.5 ±0.5		6.9±0.5
101	-	-		9.4±0.5
100	2.65 ± 0.07	15.5 ±0.5		9.6±0.5

Table 2

Temperature	Δ mm/sec	f'
77°K	0.5 ± 0.1	0.32 ± 0.06
293°K	0.46 ± 0.1	0.061 ± 0.015

Table 3

Orientation of the platelet	$f'_{77°K}$	$f'_{293°K}$
001	0.24 ± 0.05	0.054 ± 0.01
101	-	0.072 ± 0.01
100	0.36 ± 0.06	0.076 ± 0.01

1) The position of the absorption maxima was fixed more precisely by additional measurements.

2) Within the limits of experimental error, no shift of the absorption maxima was observed in spectra taken on single crystal platelets.

LITERATURE

/1/ Yu.M. Kagan, DAN, 140, 794 (1961).

/2/ Pham Zuy Xien, B.G. Shapiro, V.S. Shpinel', ZhÉTF, 42, No.3 (1962).

/3/ G.A. Bykov, Pham Zuy Xien, ZhÉTF, in press.

/4/ V.A. Bryukhanov, N.N. Delyagin, A.A. Opalenko, V.S. Shpinel', ZhÉTF, in press.

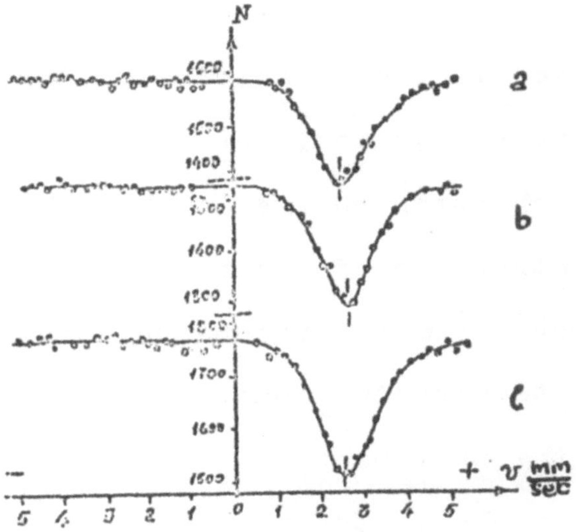

Absorption spectra taken at a temperature of 77°K for platelets of different orientations: a) 001, b) 100, c) polycrystalline. The thin vertical lines mark the position of the absorption maxima.

Translated by Robert L. Eisner

QUADRUPOLE INTERACTION AND ISOMER SHIFTS OF THE 23.8 KeV GAMMA TRANSITION OF THE Sn^{119} NUCLEUS IN TIN-ORGANIC COMPOUNDS

A. Yu. Aleksandrov, N. N. Delyagin, K.P.Mitrofanov,
L. S. Polak, V. S. Shpinel'

The first investigations /1,2/ of the spectrums of Mossbauer resonance absorption of the 23.8 KeV gamma quantums, for the Sn^{119} nucleuses in tin-organic compounds, showed that such measurements could give very valuable information as to the distribution of electron charge in the molecule and its variation with the structure of the molecule and the properties of its chemical bonds. From the resonance absorption spectrums it is possible to determine two quantities characteristic of the interaction of the nucleus with the electric field of the electrons in the molecule ; these are the isomer energy shift of the gamma transition and the constant of quadripole interaction. The magnitude of the isomer shift (denoted below by the symbol δ) is given to a first approximation by the formula $\delta = E_I - E_{II} = c(R_e^2 - R_o^2)(\psi_I^2(0) - \psi_{II}^2(0))$, where c is a constant, R_e and R_g are the effective charge distribution radiuses in the excited and ground states of the nucleus, $\psi_I(0)$ and $\psi_{II}(0)$ are the wave functions of the electrons in the region of the nucleus for the two chemical compounds, and E_I and E_{II} are the gamma transition energies in these compounds. Since for the Sn^{119}

nucleus $R_e > R_\varepsilon$ /3,4/, a positive value of δ corresponds to $\psi_I^2(0) > \psi_{II}^2(0)$. The spins in the ground and excited states of the Sn^{119} nucleus are equal to 1/2 and 3/2 respectively, so the quadrupole interaction differs from zero only for the excited state.

The quadrupole interaction is characterized by the quantity eqQ (the constant of quadrupole interaction ; e is the charge on the electron, q is the gradient of the electric field on the nucleus, Q is the quadrupole moment of the nucleus). The difference between the energies of the sub-levels with magnetic quantum numbers $\pm$ 1/2 and $\pm$ 3/2 is measured experimentally, and is found to be equal to 1/2 eqQ. This difference (the quadrupole splitting of the level) is denoted below by the symbol Δ .

A particularly interesting question is the effect on the isomer shift and the magnitude of the electric field gradient produced by replacement of one group of atoms in the molecule by others which form a series of some kind, for example in electronegativity, and so on. The Mossbauer effect can be used to gain information which could not be derived by other methods (for example, the nuclear quadripole resonance method cannot be used in the case quoted).

In the present work, resonance absorption spectrums were investigated for tin-organic compounds of the $(C_4H_9)_2 SnX_n$ type (where X represents an element or group of atoms, and n = 1 or 2), and also for certain compounds of the SnX_4 type.

DESCRIPTION OF THE EXPERIMENT

The resonance absorption spectrums were measured on two different apparatuses; in one the absorber moved at a speed which altered linearly with time /5/, and in the other, the absorber had a constant speed relative to the source. For the majority of the substances, spectrums were measured on both apparatuses. With the constant speed apparatus great accuracy could be achieved in determining the positions of the lines in the spectrums; by using with this apparatus a resonance counter /6/ which was selectively sensitive to 23.8 KeV gamma quantums, emitted without recoil, it was possible to greatly increase the relative effect of the resonance absorption (by approximately four times compared to the effect observed using a scintillation counter). The gamma quantum source was the compound SnO_2 , containing the Sn^{119m} isomer. The source thickness was 5 mg/cm^2.

In all the measurements, the source was at room temperature, and the absorber at the temperature of liquid nitrogen. Absorbers consisting of substances solid at room temperature were prepared by pressing in a mixture with beryllium oxide; liquid absorbers were prepared by filling a flat thin-walled vessel of clear plastic with the liquid. The majority of the substances investigated were specially synthesized for our measurements. Chemical analysis showed that the absorbers did not contain any impurities which would appreciably affect the shape of the resonance absorption spectrum. The thickness of absorbers varied from specimen to specimen between 30 to 100 mg/cm^2.

RESULTS OF MEASUREMENTS AND DISCUSSION

In the present work we investigated tetravalent tin compounds of the SnX_4 type in which the tin atom had a symmetrical (tetragonal) environment of X atoms or groups, and compounds of the type $(C_4H_9)_2\ SnX_n$ ($n = 1$ or 2), where the X groups or atoms were represented by halogens, oxygen, sulfur, and the acid radicals SO_4 and SO_3.

The values found for the isomer shift δ and the quadrupole splitting Δ are given in the table. Resonance absorption spectrums for compounds of the type $(C_4H_9)_2\ SnX_n$ are shown in Fig. 1 and Fig. 2.

The magnitude of the isomer shift is determined by the density of electrons in the region of the nucleus, and, as noted above, in the present case an increase in the isomer shift (i.e. an increase in the energy of the gamma transition) corresponds to an increase in $\psi^2(0)$.

It is obvious that a change in the value of $\psi^2(0)$ will primarily depend on the distribution of valence electron density between the tin atom and the other atoms present in the compound. In the case where the distribution of electrons does not have spherical symmetry, a non-zero electric field gradient can arise on the tin atom nucleus. It does not appear to be possible to assign to each element of the periodic table (or to a group of elements) a constant, quantitative parameter which would determine the distribution of electron density between neighboring atoms (or groups of atoms) in the molecule, since this distribution will undoubtedly be affected by the character of the bonding and by different collective effects. Nonetheless, within any group of molecules of similar structure, the electronegativity of the atoms can be used for this parameter /7/. A qualitative correlation has been noted previously between electronegativity and the values of δ and Δ in certain inorganic tin compounds /8/.

This tie-up with the electronegativity of the atoms present in the compound with tin is demonstrated very clearly by the values of isomer shift in the SnX_4 compounds where X is a halogen atom (Fig.3). It can be seen that δ varies linearly with the electronegativity χ, with good accuracy. A similar relationship can be detected for the values of δ and Δ in the compounds $(C_4H_9)_2SnX_2$ (where X is a halogen), but in this case the range of values of δ and Δ is comparatively small, and so the relationship is not so reliably established as as the case shown in Fig.3. It appeared of interest to try and find the relationship between the quantities determined experimentally (i.e. δ and Δ) without recourse to the concept of electronegativity. For the reasons noted above, the only relationship which could be reliably examined was that between the values of Δ for compounds of the $(C_4H_9)_2SnX_2$ type and the values of δ for compounds of the SnX_4 type (where X is a halogen atom, and the comparison is between Δ and δ for molecules containing like X atoms). In this case the relationship also turns out to be linear. It should be noted that it follows directly from this that there is a linear relationship between the magnitude of the tin atom electric field gradient in the $(C_4H_9)_2 \cdot SnX_2$ compounds and the value of δ in the corresponding SnX_4 compounds.

In the compounds $(C_4H_9)_2SnO$ and $(C_4H_9)_2SnS$ the O and S atoms are connected to the tin atoms by double bonds. From the results given in the table it can be seen that the isomer shifts and quadripole interaction constants of these compounds differ greatly from the values of the corresponding quantities in the $(C_4H_9)_2SnX_2$ series examined above (where X is a halogen). If we formally apply the relationship examined above between Δ and the electronegativity of the X atom, we would expect that the value of Δ in the compound $(C_4H_9)_2SnO$ would at least not be less than that in $(C_4H_9)_2SnCl_2$, and for the compound $(C_4H_9)_2SnS$, not less than for $(C_4H_9)_2SnI_2$. It turns out, however, that the presence of the double bond leads to a marked decrease in the gradient of the electric field on the tin atom. On the other hand, the compounds $(C_4H_9)_2SnSO_4$ and $(C_4H_9)_2SnSO_3$ show a larger quadripole interaction constant. These two compounds differ from the $(C_4H_9)_2SnX_2$ series examined above (X-halogen) in that besides the two atoms (oxygen) direcly connected to the tin atom, there are present groups of atoms (O and S) of high electronegativity. These atoms are not directly connected with the tin atom, but they exert a marked influence on the charge distribution in the molecule as a whole, and this appears to be the cause of the increase in the electric field gradient on the tin nucleus. This assumption is confirmed by the great difference between the electric field gradients on the tin nucleus for the compounds $(C_4H_9)_2SnSO_4$ and $(C_4H_9)_2SnSO_3$; the removal of one oxygen atom, not directly connected to the tin atom, leads to a marked decrease in the gradient. The induction effect of chemical bond theory can, to a certain extent, be considered as the analog of the effect observed here. The presence of atoms of high electronegativity leads to a concentration of charge in the parts of the molecule where these atoms are situated, i.e. to deformation of the molecule's electron cloud and to creation of an electric field gradient on the tin nucleus. It appears that for compounds of the $(C_4H_9)_2SnX_2$ series , the greatest deformation of the electron cloud occurs in the case where the X atom has the highest electronegativity (fluorine). This may explain the relationship between the electric field gradient and the electronegativity of the X atom.

As noted above, for compounds of the SnX_4 type the change in isomer shift on passing from SnF_4 through SnI_4 (Fig. 3) can be freely considered as due to the change in electron density on the tin atom nucleus with the electronegativity of the X atom (see also /8/). A simple extension of this interpretation to compounds of type $(C_4H_9)_2SnX_2$ meets with certain difficulties. For example, working from the numbers of highly electronegative atoms in the molecule, it might be expected that compounds of the $(C_4H_9)_2SnX_2$ series would have isomer shifts intermediate between those of SnX_4 and $Sn(C_4H_9)_4$. However, as can be seen from results in the table, δ is always much bigger for the $(C_4H_9)_2SnX_2$ compounds than for either SnX_4 or $Sn(C_4H_9)_4$. Therefore, in compounds of the type $(C_4H_9)_2SnX_2$, the presence of atoms or groups of atoms of high electronegativity in the molecule does not lead to a marked decrease in the density of the s-electrons in the region of the nucleus. This can be understood if we assume, for example, that in the $(C_4H_9)_2SnX_2$ compounds the bond between the tin atom and the halogen atoms is mainly based on the p-electrons, while the s-electron part of the wave function is mostly concentrated in the bond of the tin atom with C_4H_9 groups. In this case the halogen atoms will not exert any appreciable influence on the size of the isomer shift, and so the density of the s-electrons on the tin nucleus will be higher in the compounds $(C_4H_9)_2SnX_2$ than in either SnX_4 or $Sn(C_4H_9)_4$.

The authors thank V. I. Makarov for preparing the series of tin-organic compounds.

Institute for Petrochemical Syntheses, Academy of Sciences, USSR

Institute of Nuclear Physics, Moscow State University

TABLE 1

Values of the isomer shift δ in the energy of the 23.8 KeV gamma transition relative to the transition energy in the compound SnO_2 , and the quadripole interaction constants Δ . The values given in the table can be transformed to energy units by multiplying by 7.94. The units of δ and Δ in this case are 10^{-8}eV.

Compound	δ(mm/sec)	Δ(mm/sec)
$SnBr_4$	1.1 ± 0.1	0
$SnCl_4$	0.78 ± 0.07	0
$Sn(C_4H_9)_4$	1.3 ± 0.1	0
$(C_4H_9)_2SnF_2$	1.45 ± 0.15	3.9 ± 0.2
$(C_4H_9)_2SnCl_2$	1.6 ± 0.2	3.25 ± 0.15
$(C_4H_9)_2SnBr_2$	1.7 ± 0.15	3.15 ± 0.15
$(C_4H_9)_2SnI_2$	1.8 ± 0.15	2.9 ± 0.15
$(C_4H_9)_2SnO$	0.95 ± 0.1	2.2 ± 0.2
$(C_4H_9)_2SnS$	0.9 ± 0.2	1.9 ± 0.2
$(C_4H_9)_2SnSO_4$	1.8 ± 0.15	4.8 ± 0.15
$(C_4H_9)_2SnSO_3$	1.3 ± 0.2	4.0 ± 0.2

REFERENCES

/1/ V.A.Bryukhanov, V.I.Gol'danskii, N.N.Delyagin, E.F. Makarov, V.S.Shpinel'. ZhETF, 42, 637, (1962).

/2/ V.A.Bryukhanov, V.I.Gol'danskii, N.N.Delyagin, L.A. Korytko, E.F.Makarov, I.P.Suzdalev, V.S.Shpinel' (in the press).

/3/ V.S.Shpinel', V.A.Bryukhanov, N.N.Delyagin. ZhETF, 41, 1767 (1961)

/4/ A.J.F.Boyle, D.St.P.Bunbury, C.Edwards. Proc. Phys. Soc. 79, 416 (1962).

/5/ V.A.Bryukhanov, N.N.Delyagin, B.Zvenglinskii, S.A.Sergeev, V.S.Shpinel',PTE No.1, p.23 (1962).

/6/ N.Illarionova, K.P.Mitrofanov, V.S.Shpinel', PTE (in the press).

/7/ L.Pauling. The Nature of the Chemical Bond (Russian Translation) Goskhimizdat, 1947.

/8/ V.A.Bryukhanov, N.N.Delyagin, A.A.Opalenko, V.S.Shpinel'. ZhETF (in the press).

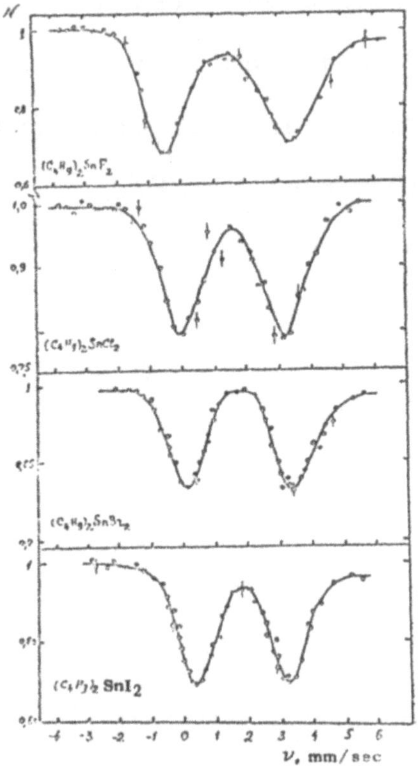

Fig.1. Resonance absorption spectrums for the $(C_4H_9)_2SnX_2$ series of compounds (X - halogen atom). The horizontal axis gives the speed of the absorber in mm/sec. A positive speed corresponds to movement of the absorber towards the source. The vertical axis shows the counting rate in arbitrary units.

These spectrums were derived on the constant-speed apparatus , using a resonance detector.

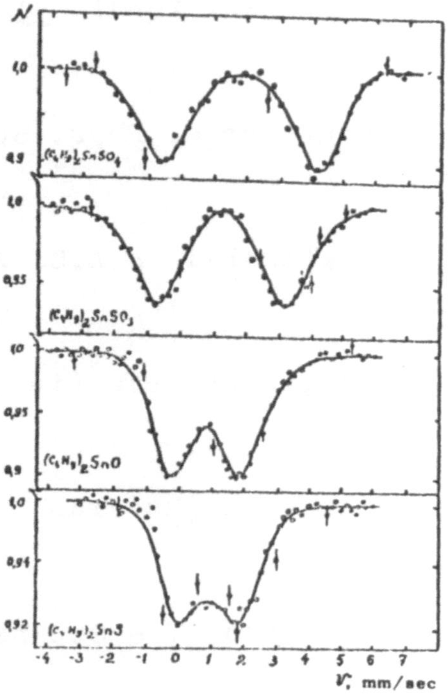

Fig.2. Resonance absorption spectrums for the compounds $(C_4H_9)_2SnO$, $(C_4H_9)_2SnS$, $(C_4H_9)_2 \cdot SnSO_4$ and $(C_4H_9)_2SnSO_3$. These spectrums were derived on the variable speed apparatus.

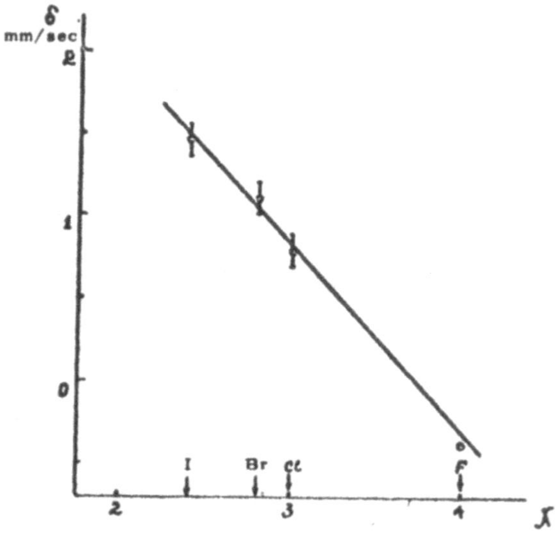

Fig.3. Isomer shift δ plotted against electronegativity of the X (halogen) atom in the SnX_4 compounds. Results for SnF_4 taken from Boyle et al /4/, and those for SnI_4 from Bryukhanov et al /8/.

Translated by Simon Lyse

INVESTIGATION OF COMPLEX IRON COMPOUNDS BY MEANS OF THE MÖSSBAUER EFFECT

A. Z. Grinkevich, G. Lizurei, E. Savitski,
T. Sénkovski

Complex iron compounds in crystalline form were investigated by means of standard apparatus for studying the fine structure of the 14.4-KeV gamma line in the absorption spectrum of iron.

Isomeric shifts and quadrupole splittings of the gamma line were measured. Results are collected in the table, where isomeric shifts are given relative to metallic iron.

The authors attempt to explain the low value of the observed isomeric shifts by assuming $3d^7 4s$ as the configuration of metallic iron in agreement with Walker and advancing a hypothesis regarding the electronic structure of the covalent bond of the iron atom in complexes.

Quadrupole splittings appearing on replacement of one $(CN)^-$ group by a ligand of another type are explained by disturbance of the spherical symmetry of the electric field.

Institute of Physics, Jagiellonian University, Crakow
Institute of Chemistry, Jagiellonian University, Crakow

Compound	Isomeric shift, mm/sec	Quadrupole splitting, mm/sec
$K_4[Fe^{II}(CN)_6]\cdot 3H_2O$	-0.09 ± 0.02	~ 0.2
$K_3[Fe^{III}(CN)_6]$	-0.13 ± 0.02	< 0.1
$Na_2[Fe^{II}(CN)_5\cdot NO]\cdot 2H_2O$	-0.24 ± 0.02	1.65 ± 0.05
$Na_3[Fe^{II}(CN)_5\cdot NO_2]$	< 0.02	1.85 ± 0.05
$Na_3[Fe^{II}(CN)_5\cdot NO_3]$	< 0.02	0.75 ± 0.05
$Na_2[Fe^{III}(CN)_5\cdot NH_3]$	-0.09 ± 0.02	1.65 ± 0.05

Translated by J. W. Loweberg, Jr.

DEPENDENCE OF THE RESONANCE ABSORPTION SPECTRUM OF γ-QUANTA ON CRYSTAL TEMPERATURE

Pham Zuy Xien and V. S. Shpinel'

Recent experiments in the resonance absorption of γ-quanta show the existence of quadrupole splitting of nuclear levels and their dependence on temperature /1/.

The resonance absorption spectra of 23.8-KeV γ-rays from Sn^{119*} at various temperatures were investigated in the present work. As source and absorber we chose polycrystalline stannic oxide (SnO_2) powder, since in this case, firstly, the probability of absorption without emission $\underline{f}'$ is sufficiently high even at elevated temperatures; secondly, the observed absorption line is appreciably broadened, which may be interpreted as a manifestation of quadrupole splitting. In our measurements the source temperature was that of liquid nitrogen; the temperature of the absorbers varied from 78°K to 645°K. Absorption spectra were taken with different thicknesses of resonance filter (from 4.5 mg/cm^2 to 20.5 mg/cm^2) with the setup described in /2/. The dependence of the half-width of the observed spectrum Γ_{obs} on the filter's thickness is shown in Fig. 1 for four temperatures. The data given clearly show the increase in width of the observed spectrum with temperature and also enable one to draw certain conclusions regarding the form of the absorption spectrum. Actually, if it is assumed in analogy with /3/ that the true absorption spectrum at the temperatures investigated has a dispersion distribution, whereas its width Γ_A exceeds the natural width of the nuclear level Γ and varies with temperature, $\underline{f}'$ can be determined from the dependence of the maximum absorption on filter thickness. The values of absorption-line broadening $\beta_A = \Gamma_A/\Gamma$ required for this purpose may be derived from measured line half-widths by extrapolating the absorption to zero thickness with allowance for self-absorption in the source. Results of calculating $\underline{f}'$ and β_A (based on the method proposed in /4/) are given in Cols. 2 and 3 of the table. On this hypothesis we obtain good agreement with experimental data for the dependence of Γ_{obs} on absorber thickness. Assuming that the true emission and absorption spectra consist of two components of the same intensity, the difference in whose energies depends on temperature, we obtain the values given in Cols. 4 and 5 for the splitting Δ and $\underline{f}'$. The dashed curve of Fig. 1 shows how Γ_{obs} depends on absorber thickness on this hypothesis. It is evident from this curve that the doublet structure of the emission and absorption lines cannot explain the rapid increase in Γ_{obs} with absorber thickness.

As was noted above, our results with regard to the observed absorption-line broadening are not in accord with the hypothesis of the doublet structure of the lines, which might result, for instance, from quadrupole interaction. The sign of the effect -- increase of line width with temperature -- also must contradict this hypothesis. Actually the electric field acting on the nucleus in the lattice may be represented in the form of the sum of the static field determined by the average positions of the atoms and the dynamic field due to thermal motion. The change in the average positions of the atoms and the amplitude of thermal lattice vibrations would necessarily lead, as in radiospectroscopic experiments in quadrupole resonance /5, 6/, to a decrease in splitting with rising temperature. In the case of certain tin compounds we observed a decrease in line width with rising temperature.

One possible explanation of the observed line broadening is connected with the effect of perturbation due to the dynamic field of the lattice, leading to natural broadening of the form of the true absorption spectrum and hence to dependence of the width of the observed spectrum on temperature. The effect of the dynamic field is considered further in /7/, the finite character of the radiation width of nuclear levels being taken into account.

The temperature shifts of the resonance absorption line in SnO_2, observed in our measurements, are shown in Fig. 2. As is well-known, the magnitude of the second-order Doppler shift is given by the equation:

$$\Delta E_\gamma = -E_\gamma \frac{\overline{V^2}}{2c^2}, \tag{1}$$

where $\overline{v^2}$ refers to the radiating nucleus.

In calculating $\overline{v^2}$ for lattices having light (O) and heavy (Sn) atoms in their composition, one can consider with sufficient accuracy that the heavy atom takes part only in acoustic vibrations. Hence the theorem of uniform energy distribution among the degrees of freedom is valid for the temperature interval under investigation. From this we find:

$$\Delta E_\gamma = -\frac{3}{2}\frac{\kappa T}{M_{Sn}c} = 3{,}50\cdot 10^{-4}\,T \quad \text{mm/sec} \qquad (2)$$

Figure 2 well confirms the linear temperature dependence of the shift, the slope of the linear part of the curve being determined by the coefficient $(3.5 \pm 0.2)\cdot 10^{-4}$ mm/sec-°K in accordance with (2).

The authors thank G. A. Bykov for a discussion of the results.

Institute of Nuclear Physics, Moscow State University

REFERENCES

/1/ V. A. Bryukhanov, N. N. Delyagin, A. A. Opalenko, V. S. Shpinel'. ZhÉTF (1962) (in press).

/2/ Pham Zuy Xien, V. G. Shapiro, V. S. Shpinel'. ZhÉTF 42, 703(1962).

/3/ D. A. Shirley, M. Kaplan, P. Axel. Phys. Rev. 123, 816 (1961).

/4/ G. A. Bykov, Pham Zuy Xien. ZhÉTF (1962) (in press).

/5/ T. Kushida, G. B. Benedek, N. Bloembergen. Phys. Rev. 104, 1364(1956).

/6/ Robert R. Hewitt. Phys. Rev. 121, 45(1961).

/7/ G. A. Bykov. Report to Present Conference /in Russian/.

/8/ R. V. Pound, G. A. Rebka. Phys. Rev. Lett. 4, 274(1960).

Table

T°K	$\underline{f}'$(I)	β_A	Δ (in $\Gamma/2$ un.)	$\underline{f}'$(II)
78	0.70	2.1	3.2	0.62
293	0.56	2.1	3.3	0.56
495	0.50	2.5	4.1	0.42
645	0.43	2.8	4.5	0.35

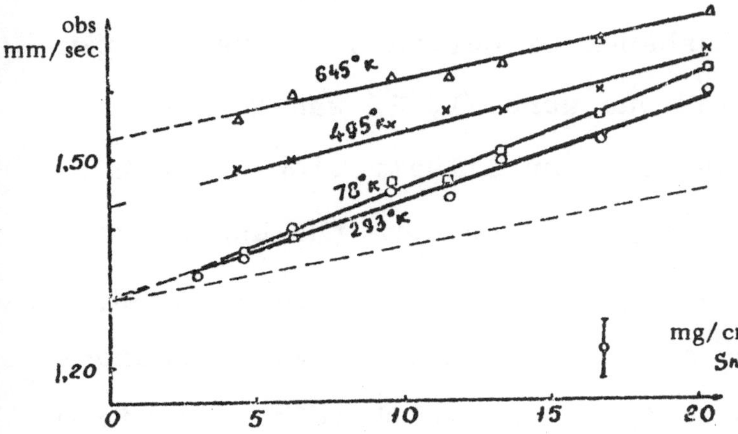

Fig. 1. Dependence of half-width of observed spectrum on filter thickness for four temperatures.

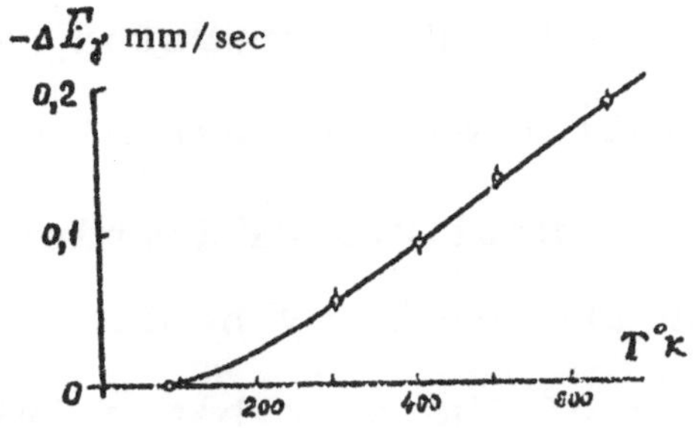

Fig. 2. Temperature shift of resonance absorption lines in SnO_2.

Translated by J. W. Loweberg, Jr.

INVESTIGATION OF THE MOSSBAUER EFFECT IN TIN COMPOUNDS

R. I. Gol'danskii, G. M. Gorodinskii, L. A. Korytko,
L. M. Krizhanskii, E. F. Makarov, I. P. Suzdalev,
V. V. Khrapov

The effect of the structure of electronic shells on the energy of nuclear gamma transitions without emission opens up a number of interesting possibilities for chemistry.

The investigation of the Mössbauer effect in quadrivalent tin compounds described here, in which our work /1, 2/ was further developed and interpreted, may serve as an illustration of such possibilities. The experiments were performed with the apparatus of the Institute of Chemical Physics, Academy of Sciences, USSR, which provided (by combining two motions -- fast and slow) uniform motion of the absorber relative to the source with velocities from zero to 15 mm/sec (at 0.07 mm/sec intervals) and from 15 to 30 mm/sec (at 0.14 mm/sec intervals). In this case the velocity was controlled continuously with an ÉPP-09 instrument. A detailed description of the apparatus will be given in a separate article. Several measurements were also carried out with the apparatus of the V. G. Khlopin Radium Institute, in which the absorber velocity varied sinusoidally with time. β-Sn and SnO_2 (the shift of the latter relative to β-Sn is -2.7 mm/sec) were used as sources of 23.8-KeV γ-quanta (Sn^{119m}). The spectra of "symmetrical" compounds of quadrivalent tin ($SnCl_4$; $SnBr_4$, SnI_4, SnO_2), in which the tin is surrounded by four identical atoms (in the case of SnO_2 each Sn atom is bonded to four O atoms, and each O atom is bonded to two Sn atoms), were investigated in the first series of experiments. In all cases singlet spectra were observed. The chemical shifts δ of lines of different absorbers relative to β-Sn, expressed in mm/sec (1 mm/sec = $7.9 \cdot 10^{-8}$ eV), are shown in Fig. 1 with respect to the electronegativity of the atoms adjacent to the tin /3/. The energy of the gamma transition obviously decreases linearly with increasing electronegativity of X in SnX_4, i.e., with increasing ionic character of the Sn--X bonds of quadrivalent tin. The chemical shift for the γ-transition in Sn^{119} nuclei is determined by the following general equation (see, e.g., /4/):

$$\delta = 1{,}6 \cdot 10^{-29} \left[|\Psi_s(0)|^2_{abs} - |\Psi_s(0)|^2_{rad} \right] \frac{\Delta R}{R} \text{ eV}, \tag{1}$$

where $|\psi_s(o)|^2$ is the density of s electrons in the region of the tin-119 nucleus and ΔR the difference between the radii of this nucleus in the excited and ground states.

In order to determine the $|\psi_s(o)|^2$ values under investigation we use data on the ionic character of Sn--Hal bonds, obtained by nuclear quadrupole resonance /5/ and in investigations of refraction and specific inductive capacitance /6/. Comparing these data with values of δ (see Fig. 1) and extrapolating to the fully ionic bond, we find that for such a bond $\delta_{ion} = -(5.6 \pm 0.5)$ mm/sec = $-(4.4 \pm 0.4) \cdot 10^{-7}$ eV. The difference between the values of $|\psi_s(o)|^2$ for fully ionized $Sn^{::}$ (completely filled n = 1, 2, 3, and 4spd shells) and β-Sn (in which $5sp^3$ electrons may be regarded as fully represented) may be considered equal to the value of

$|\psi_{5s}(o)|^2$ for one 5s electron in the field of the tin nucleus and the electrons of the shells enumerated above, since sp^3 hybridization of four orbitals in a shell does not change the value of $|\psi_s(o)|^2$ for that shell. The value of $|\psi_{5s}(o)|^2$ for one 5s electron is calculated by the Fermi-Segre formula /7/:

$$|\psi_{n,s}(0)|^2 = \frac{Z\, Z_{eff}}{\pi\, a_0^3\, n_{eff}^3}\left(1 - \frac{d\sigma}{dn}\right), \qquad (2)$$

where Z is the actual charge of the nucleus, Z_{eff} its effective charge in the outer region (with allowance for screening by electrons), a_0 the Bohr radius, n the actual and n_{eff} the effective principal quantum number, and $\sigma = n - n_{eff}$ the quantum defect. A similar calculation gives the value $1.5 \cdot 10^{26}$ I/cm^3 for $|\psi_{5s}(o)|^2$. Comparing this value with (1) and using the quantity $\delta = -(4.4 \pm 0.4) \cdot 10^{-7}$ eV, we obtain $\Delta R/R = (1.9 \pm 0.18) \cdot 10^{-4}$. Obviously, having used the given figures, one can directly determine $|\psi_{5s}(o)|^2_{abs}$ from δ -- the chemical shift relative to β-Sn. In the next series of experiments we studied the spectra of the group of organotin compounds Ph_3SnHal, where Ph is C_6H_5 and Hal is F, Cl, Br, or I. In all cases the spectra consisted of two lines, one of which lay to the right, and the other to the left, of the singlet line of $SnPh_4$ as well as $SnHal_4$. In the Ph_3SnHal spectra the center is shifted in the series of halogens in the same direction as the $SnHal_4$ singlet line but much less markedly (in the case of quadrupole splitting this may be regarded as a chemical shift).

Of special interest in the enumerated organotin compounds (as in those investigated by us earlier /1, 2/) is the pronounced doublet splitting of Mössbauer spectra, observed among "symmetrical" compounds only in SnF_4 /4/ (see Fig. 1). Here we do not reproduce the exact form of the spectra but limit ourselves to a schematic graph (Fig. 2) showing the position of each line in the doublet spectra of Ph_3SnHal and, for comparison, the positions of the $SnPh_4$ and $SnHal_4$ lines. The difference in the lengths of the two lines in Ph_3SnHal compounds illustrates the observed asymmetry of doublet splitting (the heights and widths of the two peaks were both different). This asymmetry cannot be explained by accidental orientation of the sample crystals along the direction of the γ-quanta /4/, because it persisted even when the compounds under investigation were pulverized or dissolved. As magnetic weighing and investigation of our samples in an EPR apparatus showed, the observed asymmetry of peaks could not be explained by the effect of ferromagnetic or paramagnetic impurities, either.

A special theoretical and experimental investigation of the difference between the two peaks in Mossbauer doublets, set forth in our following paper in this collection /12/, leads to the conclusion that this difference (observed for isotropic polycrystalline samples) may be caused by anisotropy of the Mössbauer effect in the corresponding monocrystals, i.e., it is due not to the orientation of the monocrystals but to their internal properties. Thus the difference between the two peaks in Mössbauer doublets by no means precludes interpretation of doublet splitting as quadrupole; moreover, it points the way to new methods for determining the properties of monocrystals from experiments with polycrystalline samples.

So, it is most natural to assume that in SnF_4 /4/, Ph_3SnHal, and other organotin compounds /1, 2/ quadrupole splitting of Mössbauer lines is observed whose value $\Delta = (1/2)eQ\partial^2V/\partial z^2$ (where Q is the quadrupole moment of the isomeric Sn^{119*} nucleus) is determined by the electric field gradient $q = \partial^2V/\partial z^2$ in the region of the Sn^{119} nucleus. Estimates of q for axial-symmetric bonds of tin with the coordination numbers 4 and 6 may be made based on the Townes method /10/. With fully covalent bonds of quadrivalent tin q = 0 both for full sp^3 hybridization and for nonhybridized s and three p bonds. The absence of a field gradient persists even in the case of fully covalent and, moreover, hybridized sp^3d^2 bonds. If with full sp^3 hybridization one of the bonds is partly ionic (fraction X), $|q| = 6.9 \cdot 10^{18}X$ V/cm^2. Similarly interpreting the data of Fig. 2 and assuming in this case that $Q = 8 \cdot 10^{-26}$ cm^2 for Sn^{119*} /4/, we find that in the series Ph_3SnHal X = 0.55 (I), 0.7 (Br, Cl), and 1 (F). However, the above interpretation is by no means the only possible one, as is obvious from the presence of quadrupole splitting in the case of SnF_4. Here the splitting may be due to d orbitals, especially since the structure of a coordination polymer with bridge-type bonds, in which each Sn atom is bonded to six fluorine atoms, is ascribed to SnF_4 /11/. With fully hybridized sp^3d^2 bonds the highest value of q corresponds to the electronic configuration 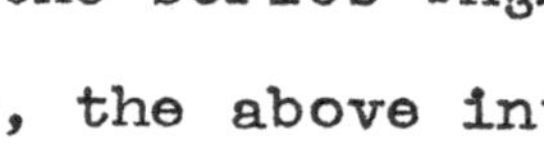 ; in this case $q \approx 9 \cdot 10^{18}$ V/cm^2 and slightly exceeds the maximum possible value for a pure p_z electron. The value $q = 4.6 \cdot 10^{18}$ V/cm^2 corresponds to the configuration 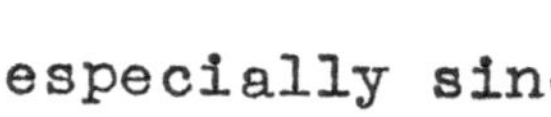, where one of the six sp^3d^2 bonds is fully ionic; this may be explained as the result of superposition of fully covalently bonded states and the two indicated configurations (predominant in the case of Ph_3SnHal), where $q = 3.8$ (I), 4.8 (Br, Cl), and 6.9 (F) $\times 10^{18}$ V/cm^2). It would be very interesting to determine directly the effective charges of the halogen and tin atoms in compounds of the type investigated in order to choose among all considered forms of interpretation of quadrupole splitting.

In conclusion let us note that when the equimolar mixture $SnPh_4 + SnI_4$ was irradiated by 1.6-MeV electrons (dose 200 megarad) in the ICP linear accelerator, we observed a marked change in form of the Mössbauer spectra, evidently due to superposition of the spectra of various disproportionation products Ph_iSnI_{4-i}. This result is encouraging from the viewpoint of applications of the Mössbauer effect not only to the study of chemical structure, but also to problems of chemical kinetics and radiation chemistry.

Institute of Chemical Physics, Academy of Sciences, USSR

REFERENCES

/1/ V. A. Bryukhanov, V. I. Gol'danskii, N. N. Delyagin, E. F. Makarov, V. S. Shpinel'. ZhÉTF 42, 637(1962).

/2/ V. A. Bryukhanov, V. I. Gol'danskii, N. N. Delyagin, L. A. Korytko, E. F. Makarov, I. P. Suzdalev, V. S. Shpinel'. ZhÉTF 43, 448(1962).

/3/ L. Pauling. Nature of the Chemical Bond /Russian translation/. Goskhimizdat, 1947. Third American edition -- 1960.

/4/ A. J. F. Boyle, D. S. P. Bunbury, C. Edwards. Proc. Phys.

Soc. 79, 416(1962).

/5/ A. L. Schawlow. J. Chem. Phys. 22, 1211(1954).

/6/ M. M. Yakshin, V. M. Ezuchevskaya, V. A. Salmenkova. Zh. neorg. khimii 6, 2425(1961).

/7/ E. Fermi and E. Segre. Z. Phys., 729(1933).

/8/ A. M. Baldin, V. I. Gol'danskii, I. L. Rozental'. Kinematics of Nuclear Reactions /in Russian/. Fizmatizdat, 1959.

/9/ Yu. M. Kagan. Doklady AN SSSR 140, 794(1961).

/10/ C. Townes and A. Schawlow. Radio Spectroscopy /Russian translation/. Inoizdat, 1959.

/11/ I. Khaiduk. Uspekhi khimii 30, 1124(1961).

/12/ V. I. Gol'danskii, S. V. Karyagin, E. F. Makarov, V. V. Khrapov. (Article in present collection).

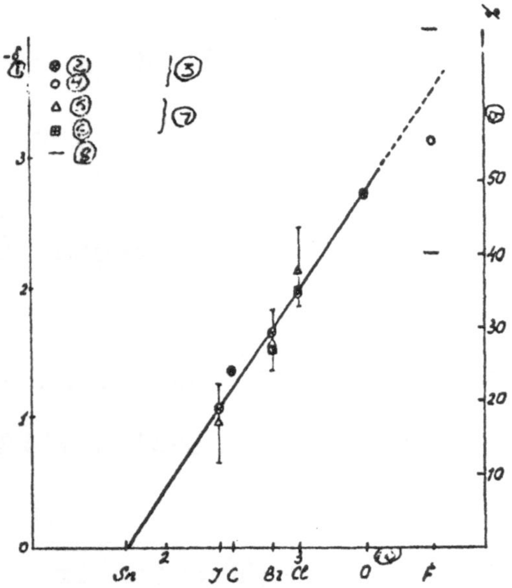

Key to Fig. 1

1. mm/sec relative to β-Sn
2. Our data
3. Mössbauer effect
4. Data of /4/
5. Data of /5/
6. Data of /6/
7. Other methods
8. Position of two lines for SnF_4 /4/ (point o plotted at center of doublet)
9. ionic character of bond
10. electronegativity

Fig. 1

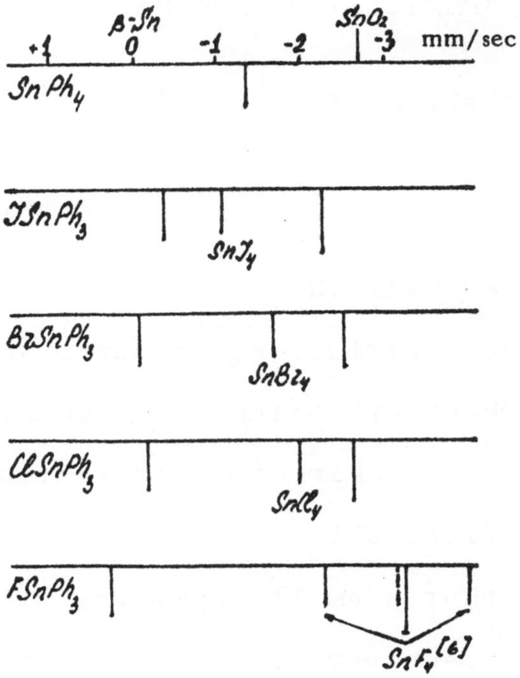

Fig. 2

Translated by J. W. Loweberg, Jr.

A POSSIBLE EXPLANATION FOR THE DIFFERENCE BETWEEN THE TWO PEAKS IN QUADRUPOLE SPLITTING OF MOSSBAUER SPECTRUMS

V. I. Gol'danskii, S. V. Karyagin, E. F. Makarov,
V.V. Khrapov

ANNOTATION

In a number of experiments, asymmetry of the two peaks has been observed in doublet splitting of Mossbauer spectrums of polycrystalline powders, making interpretation of the nature of this splitting very difficult.

It was shown that a small asymmetry will be found in quadripole splitting of Mossbauer spectrums of any polycrystalline powders (including unorientated ones), if the Mossbauer effect is anisotropic towards the corresponding single crystals. A theory was developed relating the asymmetry of the quadripole splitting peaks with the characteristic anisotropy of the Mossbauer effect towards the single crystals and with the properties of the latter.

Experimental results were derived confirming asymmetry of the two peaks in an isotropic polycrystalline specimen of triphenylchlorostannane. The degree of asymmetry for an isotropic specimen is independant of its orientation relative to the direction of the beam of gamma quantums. Partial ordering of a specimen of polycrystalline triphenylchlorostannane leads to a change in the degree of asymmetry of the two peaks, and a relationship emerges between the asymmetry and the angle of the specimen relative to the direction of the gamma quantums.

In a number of our investigations of tin-organic compounds /1,2/, quite a marked difference was observed between the two peaks in doublet splitting of Mossbauer spectrums. An example is the Mossbauer spectrum of a polycrystalline specimen of Ph_3SnCl, shown in Fig.2. Doublet asymmetry in Mossbauer spectrums has also been observed by Boyle and his colleagues for SnO /3/, and also in a number of papers /4,5,6/ on the Mossbauer effect of iron. This asymmetry for polycrystalline specimens has caused serious difficulties in interpreting this splitting of the Mossbauer spectrum as quadripole splitting. The reason for the asymmetry suggested by Boyle and his colleagues /3/, that it was due to preferred orientation of the SnO crystals in the polycrystalline specimen, was not confirmed by them in a comparison of the shapes of spectrums of SnO specimens known to be in different orientations or known to be isotropic. Besides which, in our paper /2/ it was shown that the asymmetry is completely retained on grinding or solution of the polycrystalline specimens, which at least reduces the likelihood of the explanation given in /3/. In addition, as will be seen below, for doublet asymmetry to occur it is by no means necessary for the polycrystalline specimen to have a preferred orientation, so the asymmetry must have a different origin.

It can be shown /7/ that a difference will be observed

between the two peaks in quadripole splitting of Mossbauer spectrums because of anisotropy of the Mossbauer effect towards the corresponding single crystals. In fact, for a single crystal with electric field axial symmetry lying at an angle θ to the direction of the gamma quantums, the ratio M of absorption of gamma quantums, for a nucleus of spin $\frac{1}{2}$, the excited nucleus forming sublevels of $\pm$ 3/2 (I_π) and $\pm$ 1/2 (I_δ), as can be shown from /8/, is given by

$$\frac{I_\pi(\theta)}{I_\delta(\theta)} = \frac{2\sqrt{5}\,\overline{P}_0(\cos\theta) + \overline{P}_2(\cos\theta)}{2\sqrt{5}\,\overline{P}_0(\cos\theta) - \overline{P}_2(\cos\theta)} = \frac{1+\cos^2\theta}{5/3-\cos^2\theta}\ (\mathrm{I}) \qquad (1)$$

($\overline{P}_\ell$ ($\cos\theta$) is a normalized Legendre polynomial). In polycrystalline specimens with isotropic orientation of the single crystals, and with an isotropic Mossbauer effect towards these single crystals, averaging (1) with respect to the angle gives

$$\frac{i_\pi}{i_\delta} = \frac{\int I_\pi(\cos\theta)\, d\cos\theta}{\int I_\delta(\cos\theta)\, d\cos\theta} = 1\,, \qquad (2)$$

i.e. doublet splitting of Mossbauer absorption spectrums will give two identical peaks. If however the Mossbauer effect is anisotropic towards the single crystals /9,10/, the numerator and denominator in (1) will both be multiplied by the same variable of the angle θ, $f'(\theta)$, which gives the probability of the effect. The ratio of intensities of the Mossbauer components for a polycrystalline specimen then takes on the form

$$\frac{i_\pi}{i_\delta} = \frac{\int I_\pi(\theta)\, f'(\theta) \sin\theta\, d\theta}{\int I_\delta(\theta)\, f'(\theta) \sin\theta\, d\theta} \qquad (3)$$

where in the general case, $i_\pi/i_\delta \neq 1$. In formula (3), i_π, i_δ are the values of the proportional intensities of the Mossbauer transitions $\left(\pm\frac{3}{2},\frac{3}{2}^{+}\right)\rightarrow\left(\pm\frac{1}{2},\frac{1}{2}^{+}\right)$ and $\left(\pm\frac{1}{2},\frac{3}{2}^{+}\right)\rightarrow\left(\pm\frac{1}{2},\frac{1}{2}^{+}\right)$ respectively (in the brackets are given the spin projection, the spin, and the evenness). $I_\pi(\theta), I_\sigma(\theta)$ are the full intensities of the transitions (occurring both without recoil and with transfer of lattice energy) for gamma quantums impinging at an angle θ to the crystal axis, and $f'(\theta)$ is the fraction of the transitions which occur without recoil (the factor giving the Mossbauer component of the full intensity of transition). Averaging with respect to the angle is carried out on the assumption that the single crystals in the polycrystalline specimen are orientated isotropically. Under these conditions, the ratio i_π/i_σ is, of course, independant of θ. If i_π/i_σ is found experimentally to differ from 1, and in addition a change in this ratio is observed as the orientation of the polycrystalline specimen is altered relative to the direction of the beam of gamma quantums, this indicates that there is anisotropy of the single crystals in the polycrystalline specimen, i.e. that the specimen has preferred orientation. This condition should not be inferred when dissimilar peaks are found after measurements at only one angle, although this rather unsound procedure was adopted in /3/.

The function $f'(\theta)$ can be calculated from theory /9/. It has also been found experimentally for iron in graphite /10-11/, and for single crystals of white tin /12/, and for these

$$\frac{f'(0^\circ)}{f'(90^\circ)} = 0.7 \quad \text{at} \quad T \sim 300^\circ\ K$$

Theoretical studies show that $f'(0^\circ)/f'(90^\circ)$ can vary within wide limits, according to the crystal unit cell, the elastic properties of the crystal, and the temperature /9,13/.

In the case where the crystal field has axial symmetry, the full intensities of the transitions are given by the expressions

$$I_{\pi}(\theta) = \text{const}\,[2\sqrt{5}\,\overline{P}_0(\theta) + \overline{P}_2(\theta)]$$
$$I_{\sigma}(\theta) = \text{const}\,[2\sqrt{5}\,\overline{P}_0(\theta) - \overline{P}_2(\theta)]\,, \qquad (4)$$

the function $f'(\theta)$ can also be expanded into a Legendre polynomial

$$f'(\theta) = \sum_k a_k \overline{P}_k(\theta)$$

Then the ratio of intensities of the Mossbauer components of the quadripole splitting (3) will in this case have the form

$$\frac{i_{\pi}}{i_{\sigma}} = \frac{1 + \frac{1}{2\sqrt{5}}\,\frac{a_2}{a_0}}{1 - \frac{1}{2\sqrt{5}}\,\frac{a_2}{a_0}} \qquad (5)$$

For a tetragonal lattice it can be assumed that the axis of the crystal field coincides with the c-axis of the crystal. In this case, as shown in /9/, $f'(\theta)$ takes on the form

$$f'(\theta) = \exp\{-T_1 + (T_3 - T_1)\cos^2\theta\} \quad T_1, T_3 > 0 \qquad (6)$$

where T_1, T_3 are quantities defined in /9/ which are functions of the elastic properties of the crystal along the c-axis and perpendicular to it, the temperature, the structure of the unit cell, and the energy of recoil of a free nucleus emitting a gamma quantum. By combining (6) and (4) we get

$$\frac{1}{2\sqrt{5}}\,\frac{a_2}{a_0} = \frac{1}{8}\left\{\frac{3}{\alpha^2} - 2 - \frac{6\exp[-\alpha^2]}{\alpha\,\mathrm{erf}\,\alpha}\right\} \text{ for } T_3 > T_1 \qquad (7)$$

$$\frac{1}{2\sqrt{5}}\,\frac{a_2}{a_0} = \frac{1}{8}\left\{-\frac{3}{\alpha^2} - 2 - \frac{3\exp[\alpha^2]}{\alpha F(\alpha)}\right\} \text{ for } T_3 < T_1 \qquad (8)$$

where $\alpha = \sqrt{|T_3 - T_1|}$ and $F(\alpha) = \int_0^{\alpha} e^{t^2} dt$.

A graph of i_π / i_σ against $T_3 - T_1$ is shown in Fig.1. A check shows that for $T_3 - T_1 \gg 5$

$$\frac{1}{2\sqrt{5}} \frac{a_2}{a_0} \simeq -0.25 + \frac{0.38}{\alpha^2} \quad ; \qquad \frac{i_\pi}{i_\sigma} \simeq 0.6 + \frac{0.48}{\alpha^2} \tag{9}$$

This means that the ratio of intensities for $\vec{\alpha} \ll 1$ is the same as for Mossbauer absorption for a single crystal with its c-axis lying perpendicular to the direction of the incident gamma quantums.

For $T_1 - T_3 \gg 5$, we find

$$\frac{1}{2\sqrt{5}} \frac{a_2}{a_0} \simeq 0.5 - \frac{0.38}{\alpha^2} \quad ; \qquad \frac{i_\pi}{i_\sigma} \simeq 3 - \frac{3.2}{\alpha^2} \tag{10}$$

So in this case, for $\vec{\alpha} \ll 1$, we get the same effect as when observing Mossbauer absorption for a single crystal with its c-axis lying parallel to gamma quantums being absorbed.

In those cases where the crystal field of the single crystal does not possess axial symmetry, the excited $3/2^+$ state splits into two sub-levels, which are not defined by just the spin projection quantum numbers $\pm 3/2$ and $\pm 1/2$. The spin wave functions of these two sub-levels may be shown by calculation (the matrix elements for quadripole interaction were derived by Pauling /14/) to have the form $G_\pm = N_\pm (\psi_{\frac{3}{2}} \cdot \Omega_\pm \psi_{-\frac{1}{2}}),\ G'_\pm = N_\pm (\psi_{-\frac{3}{2}} \cdot \Omega_\pm \psi_{\frac{1}{2}})$

where $N_\pm = (1+\lambda_\pm^2)^{-2}$, $\Omega_\pm = \frac{\pm f + a}{6}$, $a = \frac{3}{2} Aeq$, $b = \frac{\sqrt{3}}{2} Aeq\eta$

$$A = \frac{e\Theta}{2I(I-1)}, \qquad I = 3/2 \text{ for } Sn^{119*}, \; eq = \frac{\partial E_z}{\partial z}, \quad f = \sqrt{a^2 + b^2}$$
$$\eta = \frac{\partial E_x}{\partial x} - \frac{\partial E_y}{\partial y} \Big/ \frac{\partial E_z}{\partial z}, \qquad e\Theta = \left(II \middle| \sum_i e_i r_i^2 (3\cos^2\theta_i - 1) \middle| II\right). \text{ (II)} \tag{11}$$

$e\Theta$ is the quadripole moment of the nucleus.

$h\omega_0 \pm f$ is the resonance frequency of the gamma quantums, corresponding to excitation to the level "+" or "-" respectively. ω_0 is the transition frequency in the absence of a quadripole field.

The principal axes of the crystal field potential gradient tensor are chosen as the X, Y and Z axes. Using the theory of the S matrix /8,15/, in this case we derive, for the full probability of transitions to the "+" "-" level, the expression

$$I_{\pm}(\theta,\varphi) = const\, N_{\pm}^{2}\left\{\frac{5}{6}+\frac{\lambda_{\pm}^{2}}{2}+\left(-\frac{1}{2}+\frac{\lambda_{\pm}^{2}}{2}\right)\cos^{2}\theta-\lambda_{\pm}\sqrt{\frac{1}{3}}\sin^{2}\theta\cos^{2}\varphi\right\} \tag{12}$$

Here θ and φ are the spherical angles defining the direction of the absorbed gamma quantums in the system of the principal axes of the electric field potential gradient tensor. For the condition $\eta \to 0$ the level "+" tends to the level $(\pm 3/2, 3/2^{+})$; and the level "-" to $(\pm 1/2, 3/2^{+})$. The value of the splitting is 2a, and formula (12) is substituted in (1). Averaging (12), with the multiplier $f'(\theta,\varphi)$, with respect to all directions, we derive the intensity of the Mossbauer line for isotropic powders

$$i_{\pm} = const\, N_{\pm}^{2}\left\{\bar{A}\lambda_{\pm}^{2}+\bar{B}-\bar{C}\lambda_{\pm}\right\}, \tag{13}$$

where $\bar{A}$, $\bar{B}$ and $\bar{C}$ are the following quantities averaged with the multiplier $f'(\theta,\varphi)$:

$$A=\frac{1}{2}\left(1+\cos^{2}\theta\right),\quad B=\frac{1}{6}\left(5-3\cos^{2}\theta\right),\quad C=\frac{\cos 2\varphi}{\sqrt{3}}\left(1-\cos^{2}\theta\right) \tag{14}$$

It is not difficult to see that when $f'(\theta,\varphi) = const$, then $i_{+} = i_{-}$, and the doublet will be symmetrical. Therefore, from this point of view, asymmetry of the crystal field alone is not enough to cause asymmetry of the Mossbauer doublet in isotropic powders. The necessary and sufficient condition for doublet asymmetry is anisotropy of the Mossbauer effect in single crystals of the powder.

The ratio i_{+}/i_{-} is a function of the parameter η , and so in principle the components of the doublet may differ not only in intensity, but also in shape, as has been observed in certain cases /1,2/.

It should be emphasized that doublet asymmetry will depend on the temperature, since $f'(\theta,\gamma)$ is a function of T /9,13/.

As a preliminary experiment to test the above proposals we investigated asymmetry of Mossbauer peaks, as a function of degree of orientation of Ph_3SnCl crystals, and at various orientations of the specimen with respect to the direction of the beam of gamma quantums. The specimen of Ph_3SnCl used had a melting point of 107°C. Measurements of molecular weight by the freezing-point method, in benzene and camphor, gave values of 395 ± 21 and 357 ± 17 respectively.

In this substance, therefore, the molecules were not associated, and so the unsymmetrical doublet observed in the Mossbauer spectrum could not be blamed on the presence of associated Ph_3SnCl together with the monomer.

Measurements of the Mossbauer effect were carried out on the apparatus of the Institute of Chemical Physics, Academy of Sciences, USSR, at a temperature of 78°K. The source used was $Sn^{119m}O_2$. The specimens of Ph_3SnCl, in the form of finely dispersed, carefully ground powder, were deposited on an aluminium support (0.07 g/cm^2). Figure 2(A) shows a typical Mossbauer spectrum for this type of polycrystalline specimen. The marked asymmetry of the peaks is clearly seen. The same specimen was melted on the aluminium support and slowly cooled. The specimen then crystallized as coarsely crystalline blocks of diameter not less than 5mm. The Mossbauer spectrum of this specimen, situated perpendicular to the beam of gamma rays, is shown in Fig. 2(A). As can be seen from the graph, the character of the asymmetry is much changed in comparison to that of the isotropic polycrystalline specimen. The amplitude of the left-hand, stronger peak in Fig. 2(B) has been adjusted to that of the corresponding isotropic polycrystalline specimen. The Mossbauer spectrum of the same specimen, lying at an angle of 45° to the beam of gamma rays, is shown in Fig. 2(C). The character of the asymmetry sharply changes on the change from 90° to 45°, which is a direct indication of a certain amount of preferred orientation in our specimen. As a final confirmation of the truth of our results, after the experiments with the orientated Ph_3SnCl crystals these were ground down and carefully mixed into a powder. The Mossbauer spectrum in this case was completely identical to that measured initially for the isotropic polycrystalline specimen (Fig. 2(A)), and in this case the character of the asymmetry did not change as the orientation of the specimen was altered with respect to the beam of gamma quantums (experiments were carried out at two angles, 90° and 45°).

At the present time further experiments are being carried out to establish a quantitative connection between the theory of doublet asymmetry in Mossbauer quadripole splitting, developed above, and the determination of the parameters used in the theory to characterize the structure of the solid body. Mossbauer quadripole doublet asymmetry, due to anisotropy of the effect towards single crystals, opens the way to investigation of the structure of single crystals using results of experiments with polycrystalline specimens.

Institute of Chemical Physics, Academy of Sciences, USSR

REFERENCES

/1/ V.A.Bryukhanov, V.I.Gol'danskii, N.N.Delyagin, L.A.Korytko, E.F.Makarov, I.P.Suzdal'tsev, V.S.Shpinel'. ZhETF, 43, 448, (1962)

/2/ V.I.Gol'danskii, G.M.Gorodinskii, S.V.Karyagin, L.A.Korytko, L.M.Krizhanskii, E.F.Makarov, I.P.Suzdalev, V.V.Khrapov. DAN 147 (1962)

/3/ A.J.F.Boyle, D.St.P.Bunbury, C.E.Edwards. Proc. Phys. Soc., 79, 416 (1962)

/4/ A.Z.Grinkevich, D.S.Kul'gavchik (Poland) Report on the Working Conference on the Mossbauer Effect held 3-6 July 1962 at Dubna, USSR.

/5/ L.M.Epstein, J. Chem. Phys.,36, 2731 (1962)

/6/W.Kerler, W.Neuwirth, Z.Phys., 167, 176 (1962)

/7/ S.V.Karyagin. DAN (in press)

/8/ A.M.Baldin, V.I.Gol'danskii, A.M.Rosental'. Kinematics of Nuclear Reactions /in Russian/ M., 1959.

/9/ Yu.Kagan, DAN 140, 794 (1961)

/10/ D.E.Nagle, Report of the 2nd Conference on the Mossbauer effect. Paris, September 1961. Wiley, New York, 1962.

/11/ H.Pollak, M.DeCoser and S.Amelinckx, as /10/.

/12/ N.E.Alekseevskii, Pham Zuy Xien, V.G.Shapiro, V.S.Shpinel', ZhETF, 43, 790 (1962)

/13/ Yu.Kagan, ZhETF, 40, 312, (1961)

/14/ R.V.Pound. Phys. Rev., 79,685 (1950)

/15/ A.I.Akhiezer, V.B.Berestetskii, "Quantum Electrodynamics" /in Russian/ M., 1953.

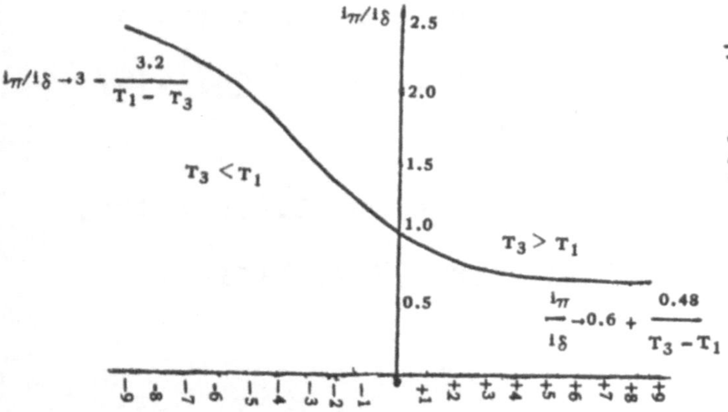

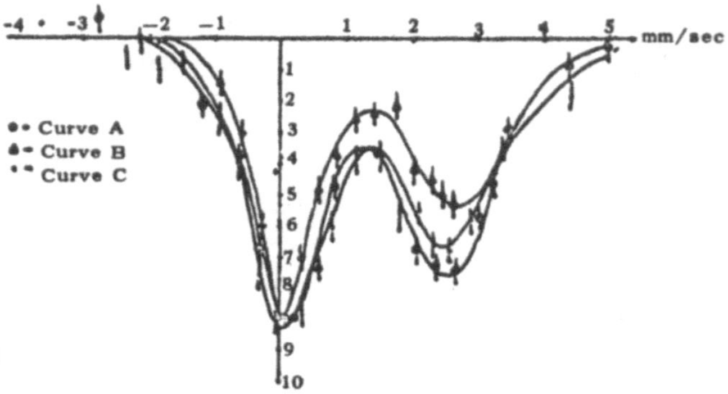

Translated by Simon Lyse